FERRARI

HYPERCARS

THE INSIDE STORY OF MARANELLO'S FASTEST, RAREST ROAD CARS

WINSTON GOODFELLOW

motorbooks

First published in 2014 by Motorbooks, an imprint of Quarto Publishing Group USA Inc., 400 First Avenue North, Suite 400, Minneapolis, MN 55401 USA

Motorbooks titles are also available at discounts in bulk quantity for industrial or sales-promotional use. For details write to Special Sales Manager at Quarto Publishing Group USA Inc., 400 First Avenue North, Suite 400, Minneapolis, MN 55401 USA.

To find out more about our books, visit us online at www.motorbooks.com.

ISBN-13: 978-0-7603-4608-2

Library of Congress Cataloging-in-Publication Data

Goodfellow, Winston, 1958-
  Ferrari hypercars : the inside story of Maranello's fastest, rarest road cars / by Winston Goodfellow.
    pages cm
  Includes bibliographical references and index.
  ISBN 978-0-7603-4608-2 (hc)
 1. Ferrari automobile--History. 2. Automobiles, Racing--Design and construction.
 3. Ferrari, s.p.a.--History. 4. Ferrari, Enzo, 1898-1988. I. Title.
  TL215.F47.G659 2014
  629.222--dc23
                              2014015103

Editor: Zack Miller
Project Managers: Elizabeth Noll and Caitlin Fultz
Senior Art Director: Brad Springer
Cover Design: Faceout Studio, Kara Davison
Book Design and Layout: Karl Laun

Printed in China
10 9 8 7 6 5 4 3 2 1

Dedication

This book is dedicated to the memory of three great names in Ferrari history:
Sergio Pininfarina, Sergio Scaglietti, and Carlo Felice Bianchi Anderloni.
   They were absolute treasures to know, and I was so very blessed to call each a friend.

CON1

8	Acknowledgments
9	Introduction
10	**Section I: The 12-Cylinder Drivetrain Era (1947–1983)**
12	Chapter 1: *Penna Bianca*
26	Chapter 2: The Hypercar Is Born (1948–1962)
58	Chapter 3: Move to the Middle (1963–1983)
90	**Section II: The Lightweight Materials Era (1984–1997)**
92	Chapter 4: A New Formula (1984–1986)
114	Chapter 5: Target: 200 mph (1987–1992)
150	Chapter 6: A Different Tact (1993–1997)
184	**Section III: The Digital Age (1998–Present)**
186	Chapter 7: Dawn of a New Era (1998–2009)
220	Chapter 8: Brave New World (2010–Today)
234	Epilogue
236	Bibliography
238	Index

Acknowledgments

I've been a student of the Italian art of making the world's fastest and most alluring road cars since the mid-1970s. My classroom has been the driver's seat and the hundreds of interviews and memorable meals shared with the men who made the cars.

In regards to this book, four "professors" stand out: the three mentioned in the Dedication (Carlo Anderloni, Sergio Pininfarina, and Sergio Scaglietti), and the last of the old guard in Maranello, Piero Ferrari. The time they so graciously spent helping to further my understanding of Ferrari's DNA is one of the highlights of my career.

Others who have yielded insight into Ferrari and its hypercars, the competitors covered in this book, and the people who made them in and outside of Marnaello are Giulio Alfieri, Anna Anderloni, Silvana Appendino, Marco Arrighi, Romano Artioli, Aurelio Bertocchi, Giotto Bizzarrini, Rosanna Bizzarrini, Aldo Brovarone, Reeves Callaway, Giordano Casarini, Giuseppe Caso, Jason Castriota, Carlo Chiti, John Clinard, Gianpaolo Dallara, Luca Dal Monte, Leonardo Fioravanti, Mauro Forghieri, Marcello Gandini, Gerolamo Gardini, Elizebetta Gardini, Paolo Garella, Franco Gavina, Francesco Gavina, Ed Gilbertson, Antonio Ghini, Jim Glickenhaus, David Gooding, Ken Gross, Davide Klutzer, John Lamm, Brandon Lawrence, John Ling, Franco Lini, Karl Ludvigsen, Franco Martinengo, Marcel Massini, Giacomo Mattioli, Nicola Materazzi, Franco Meiners, Dick Merritt, Maurizio Moncalesi, Dan Neil, Francesco Pagni, Simone Piatelli, Lorenza Pininfarina, Lorenzo Ramaciotti, Marella Rivolta, Piero Rivolta, Filippo Sapino, Sergio Scaglietti, Stefano Scaglietti, Simone Schedoni, Carroll Shelby, Paolo Stanzani, Frank Stephenson, Romolo Tavoni, Tom Tjaarda, Luigino Tommasin, Brenda Vernor, Chris Vlados, Count Giovanni Volpi, Bob Wallace, Andrea Zagato, and those who prefer to remain anonymous. To all, *grazie mille!*

Over the years, other individuals have confirmed historical information and mechanical specifications, provided hypercars to shoot and drive, assisted in photo shoots, pointed me to the right books, and much more. They are John Amette, Elena Bianchi, Alan Boe, Larry Carter, Bill Ceno, Vida Ceno, Bob Cohen, Luca Fornetti, Alfredo Garcia, Steve Kelly, Craig Jackson, Nick Loveland, Norb McNamara, Bruce Milner, Pete Racely, Mario Righini, Bruce Root, Tom Shaughnessy, Jon Shirley, Bob Smith, Matt Stone, Tom Stegman, Jack Thomas, Keith Thompson, Brandon Wang, Cullen Wetmore, Don Williams, Ted Yamashiro, Art Zafiropoulo, and those who wish to remain anonymous.

A number of archives supplemented my own, making it possible for the words *and* the photos to tell the story (the captions have been written so that if only they are read, the reader understands the plotline of Ferrari hypercar history). Those archives are the Richard Adatto Collection, the Giuseppe Busso family, John Clinard, Michael Dregni, Steve Earle, Leonardo Fioravanti, the Klementaski Collection, John Lamm, LAT Photographic, Joanne Marshall at Ferrari S.p.A., Nicola Materazzi, Giorgio Nada Editore, Pininfarina S.p.A., the REVS Institute for Automotive Research, the Sptizley-Zagari Collection, Matt Stone, Peter Vack, and Paul Vestey.

A special thanks goes to Zack Miller, my editor and publisher at Motorbooks. I was diligently working on a novel when he pitched this project, resulting in a three-week deliberation on whether to do it or not. I wanted to stay focused on fiction, but the allure of the subject was too much to pass up. I hope you, the reader, are pleased with the results.

If I have overlooked anyone here, please accept my apologies, as the omission was accidental. I'm trying to keep my publisher happy by getting him the words on time!

Introduction

The 2013 Geneva Auto Show was unlike any other.

To great fanfare Ferrari, McLaren, and Lamborghini unveiled their latest and greatest machinery (LaFerrari, P1, and Veneno, respectively) and used their own performance and "specialness" metrics to prove their car's superiority over the competition. Waiting in the wings was Porsche's 918, and in the weeks and months that followed, automotive magazines and websites had a field day trying to declare a winner before any of the cars had been driven.

In retrospect, that verbal slugfest was simply the latest skirmish in a battle that has been raging since the late 1940s. Though World War II ended in 1945, tremendous tension still existed throughout Europe, and one good way to prove your superiority over your neighbors was speed. In May of 1949 under the watchful eye of Belgium's Royal Automobile Club, Ron "Soapy" Sutton drove a Jaguar XK120 to 132.6 miles per hour. Coventry celebrated the feat in England's *Autocar* magazine two weeks later, their full-page ad declaring the XK saw "the fastest speed ever recorded by a production car."

That proclamation was the opening salvo, as other manufacturers joined the fray. Next was Spain's exclusive Pegaso marque, for in 1953 their supercharged Bisiluro set a new speed record at 151 miles per hour on the same stretch of Belgium roadside.

Since then, enthusiasts, magazines, and the automakers have generally agreed "more is better." More acceleration, more g-forces from cornering and braking, and especially a higher top speed, combined to make the ultimate definition of performance. These days, the needle has moved so far that metrics such as 0–186 miles per hour–0 and Nürburgring times define "best." Every performance-oriented manufacturer thus targets and pursues these numbers with a ruthless vengeance.

How different it was when Enzo Ferrari started making cars in 1947, and when I began following the automotive scene in the first half of the 1970s—particularly with respect to Italian cars. Back then the first thing you'd ask an owner was "How fast will she go?" Right after that came "What's she like to drive?"

That second criteria, more than any other, was what made "exotics" (and *especially* Ferraris) so mind-bendingly brilliant. They were just flat out different in how they performed and the way they made you *feel*, because they gave you an ethereal experience you couldn't find anywhere else, something that transcended far beyond merely "going fast."

Over time these machines became known as "supercars," and now their performance is so great they're dubbed "hypercars." This book focuses on these ultimate Ferraris, providing a behind-the-scenes look at their design, development, and evolution, as told by their creators. I combed through the numerous interviews I've done over the decades and read what other key people had written and said in magazine articles and books.

Then, at the very end of researching and writing, I got quite a surprise when some of the statements made by those responsible for Ferrari's latest hypercar unintentionally cast a light on how dramatically the speed game has altered, especially in the last fifteen years. A few thoughts on this emerging, game-changing trend are saved for the Epilogue, for there is an incredible, decades-long, against-all-odds tale to be told here. While innumerable Ferrari books have been written, not one has focused solely on these "fastest of the fast"—those cars as rapid or faster than what came before—and how they have continually transformed the performance landscape since the early 1950s.

We'll take care of that oversight now, so buckle up! It's one rollicking ride!

SECTION I:
THE 12-CYLINDER DRIVETRAIN ERA
(1947–1983)

"The man who creates an automobile is closer to an artist than a technician."

—Sergio Pininfarina,
former president of Pininfarina S.p.A.

In terms of one man's impact upon the thoroughbred automobile, Ferrari stands alone. In fact, he stands alone on any terms."

—automotive historian Griff Borgeson, 1975

ENZO FERRARI WAS AN AMBITIOUS, complex vanguard whose career spanned seven decades and went through three distinct phases. Admired by many, reviled by others, so great was his imprint on Italy's psyche that in his waning days, he did not travel to see the Pope.

Rather, the Pope came to Maranello.

"When you have a very high mountain to climb," coachbuilder and Ferrari confidant Sergio Pininfarina said, alluding to Enzo's thirst for success, "you cannot be normal. You must be extraordinary. You must ask your people [to do] more than the average [because] the targets you put on yourself are not normal targets, but very high targets. Therefore, working for Ferrari, or with Ferrari, was very difficult."

"An agitator of men" was how former company test driver and engineer Giotto Bizzarrini described him. Others might be more inclined to say he was a master manipulator who would do anything to get the results he desired. "In the 1950s, four or five times Ferrari offered me to be a racing driver for him," said Carroll Shelby, *Sports Illustrated*'s two-time "Driver of the Year" and father of the famous Cobra sports car. "I never did because of the politics he played with his drivers. . . . [Eugenio] Castellotti and [Luigi] Musso were good friends, and I saw what happened to them. I figured I better not get in that trap. . . ."

Juan Manuel Fangio raced for Ferrari in 1956 and was suspicious of the Formula 1 car the team fielded for him. According to the tome *Ferrari 1947–1997*, Fangio felt the mechanical breakdowns that occurred during the races were intentional, done to demonstrate that Maranello could win the F1 crown with a talented upstart (Peter Collins). To ease his mind, the then three-time world champion requested (and got) an exclusive mechanic for his car.

Such suspicion spread beyond the drivers. Giovanni Volpi's Scuderia Serenissima was likely Ferrari's largest single customer of racing machinery in the early 1960s, and the Count observed, "They didn't like clients to do well so that's where the fun was—going against them. We had to modify the cars because they were delivered 'a little bit wrong.' . . . I don't know why it was this way, for it just didn't make sense."

Then there was the famous indifference with which Ferrari would treat most clients. "[They] will wait three to eighteen months for delivery [of their car]," the prolific automotive writer Ken Purdy observed in the 1960s. "Some of them, wishing something out of the ordinary, may find it politic or necessary to go to Modena to see *Il Commendatore*. They may wait an hour for an audience. They may wait three days. . . . Sometimes desirable possessions must be paid for in more than money."

But there was a deeply human side to Ferrari that was largely invisible to the outside world. Giordano Casarini is Carrozzeria Zagato's current technical director whose career began in Ferrari's competition department in 1969. Then a young father, he built the V-12s for the 312 P, 512, and 312 PB endurance racers, learning his craft from old-time Ferrari employees. "They had an almost reverential respect for him," Casarini said. "Unbeknownst to Ferrari they called him *Penna Bianca*, referring to the noble Indian chief in the Italian version of the 1950s movie *White Feather*."

The engine builder spent every other weekend in the dyno room, including the Sunday he was not able to get a babysitter for his two-year-old daughter. He took her to the factory and, as a V-12 wailed away on the dyno, "Mr. Ferrari came in to check on how things were doing. He saw my daughter and asked somewhat brusquely, 'What's she doing here?'

"I nervously said I couldn't find someone to watch her. He replied, 'She cannot stay,' and took her away. Later I found her in his office, happily playing with a bunch of toys. Every weekend after that he would seek me out and ask, 'Where's your daughter?'"

Casarini recalled Ferrari was also generous with others: "At Christmastime he bought toys for all the engineers' children, things they couldn't afford to buy themselves, and gave them out."

Coachbuilder Sergio Scaglietti also benefited from such benevolence. Scaglietti designed and built bodies for many of Ferrari's sports-racing and road cars in the formative years, and in late 1959 was looking to move into a larger building. "I did not have the money to underwrite the expansion," he remembered. "When I wanted to buy the land, Ferrari called his banker and set an appointment. We went there together, where he co-signed for the loan."

Brenda Vernor saw Ferrari's many facets up close after moving to Italy in 1962. Then in her young twenties, she became the girlfriend of a lanky thirty-year-old named Mike Parkes, Enzo's talented development engineer and an endurance racer. She would end up working for Ferrari as a *collaboratrice* (executive assistant), a role she kept until his passing in 1988.

The Ferrari family in what is believed to be autumn 1906. Eight-year-old Enzo is on the left, his older brother Alfredo stands next to him. *Photo courtesy of Ferrari S.p.A.*

⌃ Ferrari's first step into the auto-racing world was with short-lived CMN. He's seen here competing for the first time ever in the 1919 Parma-Poggio di Berceto hill climb, where he placed eleventh overall. He also drove at the Targa Florio that year, noting in his memoirs, "It was an exciting sort of trip: We were attacked by wolves in the Abruzzi Cinquemiglia mountains. . . ." *The Spitzley Zagari collection*

⌃ Ferrari's last victory as a racing driver came at the June 1931 Bobbio-Penice hill climb. He stopped racing two months later because of "the birth of my son that changed all my life. . . ." *The Spitzley Zagari collection*

⌃ Ferrari's real talent was working behind the scenes. In 1929 he formed the Scuderia Ferrari, and the organization quickly became a force on Europe's racing scene. The Modena-based works are seen here in the early 1930s, with the team's support vehicles lining the front of the building. *The Spitzley Zagari collection*

» A rare photo from inside the Scuderia's machine shop. Two mechanics are working on a 1750 in the back, while on the right, behind another worker, a young apprentice toils away. Hiring youths was common back then, future Ferrari coachbuilder Sergio Scaglietti noting he began working at age thirteen to help support his family. *The Spitzley Zagari collection*

≽ A race that treated Ferrari very well was Italy's grueling Mille Miglia. The country pretty much came to a stop when the 1,000-mile road race occurred, spectators lining the course most of the way. This is the 1933 event in Modena, which was the first of the Scuderia's five consecutive overall wins. *The Spitzley Zagari collection*

"You really can't compare anyone to him," she said. "Ferrari was a strong man and a weak man, a very human person. Because he'd come from nothing he knew what it was like to be an ordinary mechanic who had problems. If any of his mechanics in the racing department—or anyone—had a problem with a doctor and needed a specialist, they'd go to the Old Man and he'd pick up the phone and make the appointment.

"When he arrived in the morning, he'd always say, 'Good morning,' to everybody, something you don't see bosses doing these days. He had a good sense of humor and noticed everything even if he didn't say it. If anyone had asked him what I was wearing earlier in the week, he probably would have been able to tell them."

CAREERS ONE AND TWO: RACING DRIVER AND TEAM MANAGER

In no way did Enzo's early childhood foreshadow the success and global fame he would achieve. He was the second of Alfredo and Adalgisa Ferrari's two sons, born on February 20, 1898, in Modena. In his memoirs he states he wanted to become a sports reporter or opera singer, but the seeds of destiny were planted at age ten, when his father took him to a race at the Bologna circuit. "It was watching races like that," he noted, "being close up to those cars and those heroes, being part of the yelling crowd, that whole environment that aroused my first flicker of interest in motor cars."

After the unexpected deaths of his father in 1916 and brother the following year, and a brief stint in the army shortened by illness, a lonesome and unsure Enzo traveled to Turin to find work in the auto industry. After some menial jobs his break came in 1919 in Milan when he became a test and racing driver with short-lived CMN; this lead to a seat with the Alfa Romeo works team. Moderate success, including a second overall at Sicily's grueling Targa Florio, followed.

Over the next decade Ferrari became friends with many of the era's top drivers and engineers and would become incredibly adept at operating behind the scenes. "Although I was doing well enough to justify pursuing a driving career," Ferrari observed, "I had my sights set on wider, more ambitious horizons."

He formed the Scuderia Ferrari in 1929, two years before he stopped racing. "It all began at a dinner given in honor of Alfieri Maserati and Borzacchini for having broken the 10-kilometer speed record," Alfredo Caniato, a co-founder of the Scuderia, recalled in the seminal work *Scuderia Ferrari*. "[Q]uite by coincidence I was seated next to Enzo Ferrari and Mario Tadini. Naturally the conversation centered on the sport which we all loved so much. It was then, under the auspices of Enzo Ferrari, that the foundations of the team . . . were laid."

Immediately the Scuderia was a major force in Europe's motoring scene, winning eight of the twenty-two races it entered in its first year. As his organization grew, Enzo masterfully managed the changing scene of motor racing. "In the old days," historian Griff Borgeson noted in *Ferrari, The Men, The Machines*, "there were no prima donna drivers. Even the great ones were on factory payrolls. . . . [All the drivers] knew their place relative to the marque, to the mother house."

When Alfa announced its retirement from racing in 1933, Ferrari convinced them to let him continue carrying the torch, all while dealing with the emergence of oversized but fragile egos now populating the sport. "You have to understand the period to [grasp] the enormity of what Ferrari accomplished," former Scuderia driver Rene Dreyfus recalled in *Automobile Quarterly*. "There had never been anything like the team he had, that big and so well organized. . . . If you raced for Mercedes or Auto Union or even the Alfa works team, you raced by committee . . . [Ferrari] was the whole thing [despite] the problems: the interference from the Fascists, the sensitivity of relations with Alfa, the personality problems and rivalry within the team . . . [T]here was no doubt he was the Boss—and the only boss."

⌃ One reason Ferrari's Scuderia was so successful was his masterful ability at recruiting top talent. Here, he shares a word with driving ace Achilli Varzi, who is seated in an Alfa P3. *The Spitzley Zagari collection*

NEXT PAGE:

The Scuderia first starting racing the P3 in the second half of 1933, and it quickly became its frontline racer through 1935. The famed prancing horse shield on the hood indicates this was a Scuderia car; that shield would also be used when Enzo became an independent constructor.

"[I]n the Scuderia, in the face of everything, Enzo Ferrari was relentlessly passionate about his control. The Scuderia was his dream. He was the Boss . . . the whole thing."

—Rene Dreyfus

" In a way, [the 815] pointed to the future, having lead the race for which it was designed."

— Pete Coltrin

» The first car Ferrari made from scratch was the Auto Avio Costruzioni 815. The "AAC" name came from Enzo's machine tool business. Two 815s were made in early 1940 before Italy became immersed in World War II.

⌃ The 815 badge, as seen on the nose of the lone surviving car. The "815" stood for eight cylinders, 1.5 liters.

FERRARI'S FIRST CAR—THE 815

That charisma and organizational skill carried him far, but fissures in the Alfa-Ferrari relationship formed in the second half of the 1930s when personality conflicts arose between Ferrari and Alfa's chief engineer, Wifredo Ricart, and chief executive, Ugo Gobbato. Everything culminated on January 1, 1938, when Alfa Romeo absorbed the Scuderia, and one year later the divorce was complete.

The buyout put a nice chunk of change in Enzo's pocket, but a clause in the agreement stated he couldn't compete under his own name for four years. He reorganized as Societa Auto Avio Costruzioni, a successful machine tool and aircraft engine parts manufacturing shop run from the building in Modena where the Scuderia was housed.

The gearhead in him hadn't disappeared, though. As Alberto Massimino, one of Italy's legendary engineers told journalist Pete Coltrin, "[H]is passion for racing cars had not abated, nor was it on the wane." All it needed was a spark, and that it got when Antonio Ascari (son of racing driver and Ferrari friend Alberto Ascari) and Marquis Lotario Rangoni Macchiavelli di Modena paid him a visit, asking him to build them a pair of sports cars.

Ferrari surrounded himself with serious talent for the endeavor—the aforementioned Massimino, engineer Vittorio Bellentani, test driver Enrico Nardi (who would later gain fame for his steering wheels), and Felice Bianchi Anderloni of Carrozzeria Touring. In four months they created the 815, a two-seat roadster with lightweight aluminum coachwork, and a 1.5-liter eight-cylinder engine. Both cars competed in the 1940 Gran Premio Brescia della Mille Miglia where they dominated their class, but retired after one ran as high as tenth overall.

Six weeks later Italy declared itself an Axis member and became engulfed in World War II. According to *Enzo Ferrari's Secret War* by David Manton and research done by Ferrari historian Luca Dal Monte, Enzo lived a perilous life for the next several years, doing a masterful balancing act of caring for his workers while keeping the occupying Nazis, the Socialists, and the Fascists at bay.

⌃ An 815 undergoing testing in 1940. Note the emptiness of the roads and surrounding countryside. Back then Italy's Reggio Emilia region was agricultural and poor; now it is one of the most prosperous areas in all of Europe. *The Spitzley Zagari collection*

This is the only known photo of Enzo Ferrari taking the first Ferrari (sans coachwork, which hadn't been made yet) out for its inaugural drive. Everything that has followed can be traced back to this moment on March 12, 1947. *Giuseppe Busso*

He also assisted the Partisans' resistance efforts. Manton writes that several former employees recalled after Ferrari moved the company to Maranello in 1943, the works were used at night to repair and recommission Partisan weapons in "workshop areas . . . with no lights showing [in case] anybody walked past the factory late at night." The final two years of conflict were particularly tense. Ferrari received death threats and did a smuggling run where he personally drove a prominent pro-Partisan politician to safety.

Through it all, he harbored a dream of returning to competition. That moment came in spring 1945, when the Germans surrendered to the Allies. Italy's infrastructure and industrial capability was decimated, but "I was not unprepared for the end of the war," Ferrari noted in his memoirs. As he wound down his machine tool business, he placed a call to a most welcoming recipient in Milan.

Gioachino Colombo was then forty-two years old, a gifted draftsman and former head of Alfa's technical department who worked with Ferrari in the last years of the Scuderia. Then under suspicion of being a Fascist supporter, "For me," Colombo wrote in *Origins of the Ferrari Legend*, "[Ferrari's call] was something that could obliterate in one stroke those five years of war, bombardments, and sufferings. . . ."

THE FINAL PHASE: PROPER MOTORCAR CONSTRUCTOR

Colombo recalled the 100-plus mile sojourn to Modena was "something of an adventure." He and traveling companion Enrico Nardi encountered "bridges . . . still in ruins," and ferry rides across rivers that were nerve wracking. "After long hours of waiting under the broiling sun, your turn finally came, and onto the barge you drove, balancing precariously on the runway of the landing stage."

The trip was worth it. Ferrari and Colombo quickly agreed on the car's mechanical specifications, especially the powerplant. "I had the opportunity of observing the new Packard twelve-cylinder on the splendid vehicles belonging to high-ranking American officers," Ferrari recalled, so his machines would also use V-12s.

Enzo's first car was the "125," the model designation coming from the size of one cylinder in cubic centimeters. Three Weber carburetors would feed the 1.5-liter all-aluminum SOHC engine, and horsepower quotes ranged from 72 to 118 horsepower, depending on if it was supercharged or not. The gearbox was a five-speed, which was rare at the time. Other mechanical highlights included a chassis made of oval tubes with center cross bracing for stiffness, a suspension with unequal length wishbones and hydraulic shocks, and an anti-roll bar in back.

At his apartment in Milan, Colombo toiled away on the 125 until November 1945, when Alfa came calling. They wanted him to return, and he recommended Ferrari use Giuseppe Busso, a skilled thirty-two-year-old technician with Alfa roots, to continue the work. In *Ferrari Tipo 166*, Busso noted that he "had to be ruthless in order to pull the 125 through its childhood illnesses"; those maladies included a lack of high-grade materials and badly made components from suppliers.

To illustrate how acute material shortages were, restorer and Pebble Beach judge John Ling was a nineteen-year-old working at southern California's S&A Italia Sportscars in the 1960s when shop owner Sal di Natale waved him over to an elevated lift. "This is how we made our cars in the 1940s," the former Maserati racing mechanic said, pointing to a Cinzano aperitif sign that was part of a Ferrari's coachwork. "We got metal anywhere we could."

⌃ May 1947 saw the Ferrari marque compete for the first time with a *tipo* 125 S (front row, far left). Driver Franco Cortese ran away from the field, only to retire with fuel pump problems. Two weeks later, at the Rome Grand Prix, Cortese scored the first of the marque's countless victories. *The Spitzley Zagari collection*

That first 125 no longer exists, so from December 1990 to May 1991 Ferrari's craftsmen created a replica (chassis 90125) of that car, using the original blueprints to ensure exactness. Dino Cognolato made the replica's body; he was a key figure in producing the F40's carbon-fiber coachwork.

On September 26, 1946, Ferrari's V-12 fired up for the first time. Four months later a proud Enzo drove his rolling chassis. May 11, 1947, was the company's coming out party when it entered two 125s in the Circuito di Piacenza race. Only Franco Cortese's car would start, and from the front row he jumped out to a large lead before a fuel pump failure prevented him from finishing.

Two weeks later at the Rome Grand Prix, Cortese scored Ferrari's first victory. "We would race every Sunday to test the car," he recalled in *Ferrari Tipo 166*. "The others were mainly Maserati, but we were superior. . . . I would say exceptional for that period. . . . We were used to normal four- and six-cylinder motors [but the] twelve-cylinder was like an electric motor. It would spin freely, so you had to be careful."

Cortese's "spin freely" remark highlights the first cornerstone of the Ferrari legend, one that played *the* key role during the first three decades of Maranello's hypercar production—a stellar, charismatic powerplant. Yet even Enzo Ferrari would soon realize an engine alone didn't make for a memorable, sellable car. Onlookers and potential customers had to be able to identify and then recall what it was that went so fast.

Which is how the second major element of the hypercar came into being.

« The heart of that first car—and the Ferrari legend—was a stupendous engine. Depending upon the state of tune in 1947–48, the 1,496cc V-12 made 72 to 120 horsepower.

« The tach shows what made the 125 S engine (and all subsequent Ferrari powerplants) so special: an unparalleled ability to run at high revs.

THE HYPERCAR IS BORN

(1948–1962)

» This Allemano-bodied 166 S (chassis 003 S) was a one-off, and it won the 1948 Mille Miglia. Driving it to victory was Clemente Biondetti. Here it is at the start of the race. *Giorgio Nada Editore*

≈ The model that gave the Ferrari marque a distinct visual identity and worldwide recognition was Carrozzeria Touring's seminal 166 MM barchetta. Here is chassis 0008 M on its way to winning Le Mans in 1949. *The Klementaski Collection*

A CORE PHILOSOPHY of Ferrari's first era hypercars is found in an observation Enzo made of the 125: "I always paid more attention to the engine than the chassis, convinced that engine power accounted for over 50 percent of any success at the track."

Giuseppe Busso thus followed that tact. "Work became frantic after the first race in Piacenza," he recalled in *Ferrari Tipo 166*. "I intended on increasing the [engine] cubic capacity to 2 liters. We got there by stages, first with the 159, . . . and later with the 166, which gave its first coughs at the end of November [1947]."

TURNING POINT: THE 166 MM

It was the 166 model that put Ferrari on the path to legendary status as the preeminent producer of hypercars. The 166 had the same basic mechanical layout as the 125 (and 159): tubular chassis, independent suspension up front and rigid axle in back, five-speed gearbox, and light alloy drum brakes.

Several 166 models were offered: Spyder Corsa, Inter, Sport, MM (Mille Miglia), and Formula 2, with varying wheelbases (2,200–2,620mm), and power outputs (90–160 horsepower). They carried a variety of coachwork, depending on the car's purpose: coupes, berlinettas (fastbacks), homely spyders with cycle fenders, and barchettas (two-seat roadsters with minimal amenities).

If all this sounds confusing, there's a simple reason why—it was. For instance, at the 1948 Mille Miglia, the overall winning 166 (chassis 003 S) was a one-off Allemano coupe that looked nothing like the other 166s competing. Ferrari recognized this shortcoming and turned to Felice Bianchi Anderloni for help. Anderloni's Carrozzeria Touring's coachwork had graced many of Scuderia Ferrari and Alfa's most successful prewar road and racecars, as well as Enzo's first car, the 815.

No sooner had dialogue begun than the effort was dealt a major blow. Felice suddenly passed away on June 3 that year

At the time of the 1948 Mille Miglia victory, the Ferrari marque had an identity problem. Enzo's cars sported a number of ungraceful, indistinctive body styles such as the MM-winning Allemano coupe and the cycle-fendered coachwork fitted on a number of cars such as chassis 010I (pictured).

⌃ The man Ferrari tasked with giving his cars the visual chutzpa needed was Carrozzeria Touring's affable and talented styling director, Carlo Felice Bianchi Anderloni. "In my mind," Anderloni recalled of his work at the time, "it was absolutely necessary to give Ferrari his own emblem, an identity. The result was the 166 MM barchetta."

from an apparent heart attack, the unexpected death making his son, Carlo, Touring's new design director. Then age thirty-two, the effervescent Carlo Anderloni said he was "born in the middle of cars, so it was *always* interesting!" Case in point: Felice drove to their apartment for lunch each workday, and Carlo would sit at a window, waiting for his father to park on the street below so he could "judge the car's form and lines." He also frequented the Touring works on Sundays, running his hands over unfinished metal forms while Felice conversed with the handful of workers there.

Carlo had worked in the carrozzeria for five years when his father died, so he well knew the weight he now carried. Touring was Italy's most influential coachbuilder, Ferrari was a big name, and Anderloni suspected everyone would question his ability to fill his father's massive shoes. After consulting with business partner Gaetano Ponzoni, Carlo called Enzo to tell him the news, then nervously traveled to Modena to meet Ferrari and the recently returned Gioachino Colombo.

Once there, Anderloni got a surprise: "Ferrari wanted several cars for 1949's Mille Miglia, all looking the same. He wanted to appear like a proper motorcar constructor, with an appearance that when someone saw his cars they would say, 'That's a Ferrari.' So he ordered a number of them."

A rejuvenated Anderloni returned to Milan, determined to give the fledgling marque an identity: "His car was new, with a twelve-cylinder engine that was new, so the body should also have a new appearance—not extravagant, but technical, something that was fresh." Through his grieving, Carlo's mind frequently raced on the assignment, and more than once he awakened during the night to make notes or sketches on a pad of paper on the bedside table.

Three months later Touring's seminal shape debuted at the 1948 Turin Auto Show, the 166 MM roadster marking a turning point in Ferrari history. Giovanni Canestrini was one of Italy's most influential automotive journalists and personalities, and he was dumbfounded when he saw Ferrari chassis 0002 M on display. "I can't judge that as a motorcar," Anderloni recalled the journalist blurting out, "for its shape is completely new. It's really a *barchetta* [little boat]."

The name stuck, and a handsome berlinetta soon followed. Open and closed 166 MMs went on a tear throughout 1949, winning the Mille Miglia and Le Mans (both in 0008 M), and eighteen other races. Identical road cars starred on the show circuit, *Autocar*'s Paris Show coverage capturing the barchetta's essence when they noted, "For sheer economy of style there's little to challenge this Superleggera sports two-seater. . . ."

The success of the 166 barchettas and berlinettas served as a template for every Ferrari hypercar model and brought in the second and third ingredients: Take race-proven technology, clothe it with a svelte, lightweight suit that looked as good or better than anything else on the road, and sell it in extremely limited numbers.

340 AMERICA: THE HYPERCAR PRECURSOR

By 1949, Europe was in the midst of U.S. Secretary of State George Marshall's plan to revitalize the continent. From 1948 through 1952, sixteen nations received $17 billion in economic assistance. Money flooded into Italy and elsewhere and "every street number became a building site," coachbuilder Battista Pininfarina noted in his autobiography. "People came up with new trades and new occupations for themselves."

As Italy's economy strengthened, a number of regions such as Umbria, Calabria, and the coastal town of Senigallia hosted races. Italians embraced their passion for motorsport, style, and speed, encouraging numerous gentleman racers to come out of the woodwork. Favored marques included Ferrari, crosstown rival Maserati, small-displacement specialist Stanguellini, and OSCA.

Demand also existed on the opposite side of the Atlantic. The Sports Car Club of America (SCCA) had formed in 1944 on the East Coast, and the Sports Car Club of California came into existence three years later. Both were staging races by the late 1940s, with the SCCA's successful race at Bridgehampton in 1948 opening the import

⌃ Driving 0008 M to victory at Le Mans was Luigi Chinetti, seen here at a 1984 Ferrari Owners' Club meet in Monterey, California. He'd won the race twice before with Alfa Romeos, and as Ferrari's American importer, had much influence during the first three decades of the marque's existence.

« At the time of Chinetti's Le Mans victory, Ferrari was a small constructor focused primarily on racecar production. This is the competition department in September 1950; in the foreground is a 125 C, while over to the left the first 375 F1 is under construction. The latter's powerplant would play a major role in Ferrari's first hypercars. *The Mailander collection at the Revs Institute for Automotive Research*

» September 1950, and the courtyard of the Ferrari factory looks quite serene. That year, the works would produce barely two cars a month. *The Mailander collection at the Revs Institute for Automotive Research*

≈ The first Ferrari model to edge toward hypercar-type performance was the 340 America. Its top speed approached 150 miles per hour, and this Touring barchetta marked the America's world debut at the 1950 Paris Auto Show. *The Mailander collection at the Revs Institute for Automotive Research*

floodgates. Foreign car dealerships sprang up to cater to increasing demand for sports cars.

Italian-born Luigi Chinetti led the Ferrari charge in America. In his early forties at the time, Chinetti was born in Milan, the son of a mechanical engineer. After attending college for two years he went to work with his father and began racing in 1928.

Much of Chinetti's career was defined by being in the right place at the right time. By 1932 he was part of Alfa's competition department, where his and Ferrari's paths crossed. Luigi went to Le Mans for the first time that year, taking the overall win in an Alfa 8C. His second victory came two years later in another 8C. By now he was strongly anti-Fascist, so he left Italy and settled in Paris.

In 1940, just weeks before war engulfed all of Europe, Chinetti came to America as part of Lucy Schell's Maserati Indianapolis team. He stayed on in the United States, obtaining work as a mechanic at Rolls-Royce distributor J. S. Inskip and gained his citizenship in 1946. That winter he visited Enzo in Modena and told him numerous wealthy Americans would purchase his nascent car.

Luigi would become Ferrari's U.S. importer and eventually sell 25–30 percent of Maranello's annual production. His first Ferrari sales occurred in 1948 to famed sportsman Briggs Cunningham and radio station owner Tommy Lee. Enzo acknowledged the importance of the U.S. market in 1950 when he named his most powerful road car ever the 340 America.

The model's roots can be traced to Ferrari's Formula 1 efforts, an early example of how Ferrari applied race technology to road cars and a clear precedent for later hypercars. Before the war, F1 was known as Formula A, and regulations devised in 1939 stated that beginning in 1941 cars could have a 1.5-liter supercharged engine, or up to 4.5 liters naturally aspirated. Those rules were implemented when racing resumed after the war, and Ferrari made its official F1 debut at the Italian Grand Prix in September 1948. The 125 F1 featured a 1,497cc V-12 with a single-stage Roots supercharger and made around 230 horsepower. That wasn't nearly enough to contend with Alfa's 158 and its 300 horsepower, supercharged eight-cylinder engine, but a mold had been cast.

A year later Gioachino Colombo left Ferrari and was replaced by his former assistant Aurelio Lampredi. Then in his forties, Lampredi had previously worked at aircraft manufacturers Piaggio and Reggiane-Caproni and would be the first of many designers and engineers to apply aircraft technology to Ferrari's cars. "I came from a job where the man in the sky is the one with the engine," he noted in *Ferrari Tipo 166*. "If the engine stopped he would come down like a stone."

Prodigious, reliable, efficient horsepower ruled Lampredi's designs. He'd noticed the supercharged Alfa's thirst, so he set Ferrari on a new course by devising what became known as the "long-block" engine. His design increased the spacing between the cylinders' axis, allowing Ferrari's V-12s to grow from 2 to nearly 5 liters. He also applied aircraft technology by screwing the cylinder liners into the heads and eliminating the need for head gaskets. The head and barrel assembly is then attached to the crankcase. This made the heads one with the block, allowing for compression ratios to increase, which boosted engine power. The first displacement bump was to 3.3 liters in the short-lived 275 S, then 4.1 liters for the 340 America. The new model debuted at the Paris Auto Show. The Touring-bodied car looked like a slightly larger 166 barchetta. The wheelbase increased 2,200mm to 2,420mm, and the front and rear track grew by 28 and 50mm, respectively, to 1,278 and 1,250. The gearbox stayed a five-speed, and the front suspension remained independent with double wishbones and transverse leaf springs. The rear had a rigid axle and semi-elliptic leaf springs. Curb weight was listed at 900 kilograms (2,080 pounds), so the 220 horsepower engine didn't have much mass to move.

Autosport tested a 340 America barchetta for its May 4, 1951, issue, and the results were stunning. Driving the car was Giannino Marzotto, winner of the 1950 Mille Miglia,

« The close parallel between Ferrari's road and racecars is highlighted here with 340 America chassis 0032 MT at the 4:16 a.m. start in the 1951 Mille Miglia. Outside the air intake just above the grille, it looks identical to the Paris Show car. *The Mailander collection at the Revs Institute for Automotive Research collection*

≍ The 340 America that won the 1951 Mille Miglia was this one-off Vignale-bodied car, chassis 0082 A. That victory, with the punched-in nose, is a testament to the robustness of Ferrari's mechanicals and the car's driver, Gigi Villoresi. *Giorgio Nada Editore*

» Key to 0082 A's performance was its 4,101cc V-12. While its 220 horsepower may not seem like much today, back then the typical American car made barely half that and weighed considerably more. And European cars had even less horsepower.

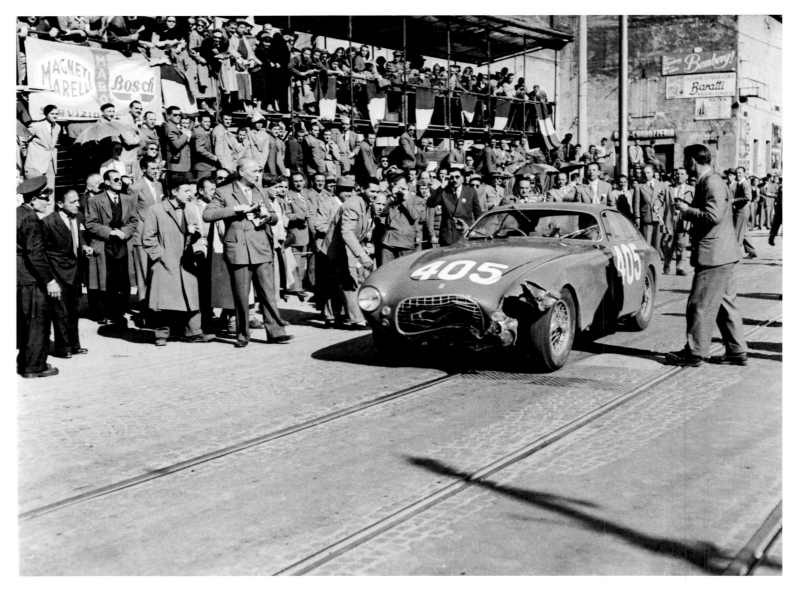

340 AMERICA

Year(s) made: 1950–1952
Total number produced: 23 (the hypercar precursor)
Powertrain: 4,101cc SOHC V-12, 220 hp @ 6,000 rpm; five-speed transmission
Weight: 1,763 lbs*
Price when new: 5,000,000 lire ($8,000 U.S., f.o.b. Italy)

PERFORMANCE:

0–60 mph: n/a
0–100 mph: n/a
1/4 mile: 13.8 seconds
Top speed: 150+ mph

Road tested in: *Autosport*, May 4, 1951; * Touring barchetta in *Road & Track*, December 1950
Main competitors: Maserati A6, Jaguar XK120, Aston Martin DB2

≈ The early 1950s heralded the explosion of custom coachwork on Ferrari chassis, and Carrozzeria Vignale was the main beneficiary through 1953. Today, 0082 A is completely restored and still a beast to drive—as this book's author can testify to.

and he recorded a quarter-mile time of 13.8 seconds. *Autosport* called the acceleration "astounding," found the handling to be "a revelation," and the suspension "beyond criticism . . . iron(ing) out uneven road surfaces to an unbelievable degree." With a top speed estimated at "over 150 mph," they concluded the 340 America "is undoubtedly the fastest sports car ever placed on the world's markets. . . ."

To put that level of performance in contemporary context, consider *Road & Track*'s article from the same month. Tests of a Studebaker Land Cruiser (akin to today's Honda Accord coupe) netted a 21.1 second quarter mile and a top speed of only 95 miles per hour. Much faster but still far behind the 340 was Jaguar's XK120, which covered the quarter in 18.3 seconds, and hit 123 miles per hour flat out.

The Importance of Design

Not only did the 340 go faster than anything else, it looked the part. "For my father," current company vice-chairman Piero Ferrari observed, "coachwork was very, very important. He was an engine-oriented person, so that was first. But second was coachwork, for style was [critical]. . . . I remember many times my father . . . watching a prototype that wasn't painted, just sitting there in bare aluminum. He'd say, 'I don't like this radius here. Make it bigger, make it sharper.'"

Touring's barchettas and berlinettas were undeniable game changers, but the coachbuilder didn't monopolize 340 America production. In the early 1950s Carrozzeria Ghia was Italy's third oldest coachbuilder, run by commercial director Luigi Segre and chief designer Mario Boano. They produced a small number of Ferraris, including five 340

⌃ One of the most spectacular 340 Americas was this rakish Vignale Spyder, chassis 0140 A. It was built in 1952 and went to American importer Luigi Chinetti. *John Lamm*

« Carrozzeria Ghia in Turin made just four 340 Americas. This berlinetta is chassis 0148 A, the second to the last of the series. A (nearly identical car chassis 0150 A) finished fifth overall at the 1952 2,000-mile Mexican road race, the Carrera Panamericana.

ⱴ The man responsible for making the 340 Americas and other Ferraris go faster was Aurelio Lampredi, seen here in 1951 at Monza with Enzo Ferrari. Among the many tricks Ferrari's engineering chief relied on was one made famous by American hot rodders—stuffing ever-bigger engines under the hood. *The Mailander collection at the Revs Institute for Automotive Research collection*

» In the early 1950s the Ferrari with the largest engine was the 375 F1. It's 4,494cc V-12 started with 350 horsepower, but jumped to 380 for the 1951 season, thanks to the addition of twin plug heads. At the 1951 German Grand Prix, future star driver Stirling Moss snaps a shot of Jose Froilan Gonzales. *The Mailander collection at the Revs Institute for Automotive Research collection*

Americas. One berlinetta (chassis 0150 A) that looked much like a road car finished fifth overall at the famous Mexico Panamericana road race in 1952.

Far more prolific was Carrozzeria Vignale. Alfredo Vignale learned panel beating before the war at Stablimenti Farina and Pinin Farina and branched into car design at age thirty-three, when he formed his company in 1946. "He'd sketch cars on the floor," said Francesco Gavina, a Vignale work associate and friend. "He really wasn't a designer, so the sketches were a bit crude and the workers would have to interpret them, to get [the coachwork] to what Vignale wanted it to look like."

Vignale's real talent was what he could do with aluminum. "He was an artist with the hammer," Gavina said, and the master panel beater became Italy's most prolific coachbuilder in the early 1950s, when he teamed with stylist Giovanni Michelotti. Then barely thirty years old, Michelotti "was 100 percent a gentleman, very shy" and had so much innate talent that Gavina says, "When I'd ask him, 'How can you design like this?' he'd answer, 'I don't know.'

"It's difficult to describe how big he was in stature [for] he could design with the eyes closed. . . . When he got the job with Triumph, he boarded the plane [to England] with an empty sketchbook. When he arrived, it was full. *This* was Michelotti."

Much like Vignale's workers "translated" his floor sketches, Vignale did the same with Michelotti's stylized renderings. "They had great respect for each other," Gavina noted. "The character of Vignale was really strong, and he used to say, 'What I do is correct.' But going against Michelotti was very difficult because he was so talented."

Of the nine 340 Americas Vignale and Michelotti produced, two stand out. Chassis 0082 A was the coachbuilder's first 340 America, and this smooth jellybean of a berlinetta won the 1951 Mille Miglia despite Gigi Villoresi losing the use of second and third gear. The second was chassis 0140 A, a lithe spyder that screamed speed when standing still.

These two 340s barely scratched the surface of Vignale and Michelotti's accomplishments. In *Ferrari by Vignale*, author Marcel Massini notes the carrozzeria bodied 156 Ferraris between 1950 and 1954. Using that figure, 72 percent of all the cars Enzo made during those years carried Vignale coachwork.

FRAMEWORK FOR THE FIRST HYPERCAR

As impressive as the 340 America was, Ferrari's first true hypercar came in 1954. There's so much to say about the 375 MM that it deserves its own book, but its lusty V-12 dates back to the 1950 Formula 1 effort, when a 375 F1 debuted at the Italian Grand Prix at Monza in September 1950. Lampredi had increased the engine bore to 80mm and the stroke to 74.5mm, for a total capacity of 4,494cc. Compression jumped from 8:1 to 11:1, and three larger Weber carburetors (42 DCF) were used. With output now 350 horsepower, at the 375 F1's first race it finished second to Alfa's 159.

Lampredi further developed the engine by adding twin-plug heads, bumping power to 380. In July 1951 Enzo finally claimed the prize that eluded him: a victory over Alfa when Froilan Gonzales' 375 F1 won the Silverstone Grand Prix. Ferraris won the next two races and narrowly lost the title when Enzo made a bad tire choice for the last race. Alfa won the battle and the championship but saw the writing on the wall and announced its withdrawal from competition.

Two years later the 4.5-liter V-12 found its way into Maranello's sports racers. As Piero Ferrari noted, "The real legend of Ferrari was created in endurance racing not Formula 1," and a big part of that legacy was the 375 MM.

At the beginning of the 1953 season—the first year of the World Sports Car Championship—Ferrari's big gun was the 340 MM; within months the more powerful engine was shoehorned into three 340 chassis (the 0318 AM, 0320 AM, and 0322 AM). Such "hybrids" were typical then, where constant development preceded production of

≳ It was only a matter of time before the 375 engine found its way into Maranello's endurance racing machines. In the early months of 1953, three 340 MMs received the engine. The last was 0322 AM, shown here under hard acceleration at the Monterey Historic Races.

375 MM

Year(s) made: 1953–1955
Total number produced: 24
Hypercars: 10 (0368 AM, 0378 AM, 0402 AM, 0416 AM, 0450 AM, 0456 AM, 0460 AM, 0472 AM, 0476 AM, 0490 AM)
Powertrain: 4,494cc and 4,522cc SOHC V-12, 340 hp @ 7,000 rpm; four-speed transmission
Weight: 2,400 lbs (berlinetta); 2,100 lbs (spyder)
Price when new: n/a

PERFORMANCE:
0–60 mph: n/a
0–100 mph: n/a
1/4 mile: n/a
Top speed: 172 mph
Road tested in: *Peter Coltrin racing in color*, reported top speed of a "brand new" 375 MM spyder at the Bonneville Salt Flats in April 1954
Main competitors: Mercedes-Benz Uhlenhaut Coupe; Pegaso Berlinetta Cupola, Spyder Touring Competizione, Spyder ENASA Rabassada; Jaguar D-type

the actual Ferrari model. The 340/375 MMs won at Spa, Senegallia, and Pescara, helping Ferrari secure the first of its seven endurance titles through 1961.

The "production" 375 MM debuted in November 1953, and two engines were used in the cars. The first was the F1-derived *tipo* 102 4,494cc V-12, as found in the 340/375 hybrids. The other was an enlarged version of the 340's 4.1, the *tipo* 108 4,522cc V-12. Both retained classic Lampredi engineering touches: heads in unit with the block, SOHC, and more. Carburetion was three dual-throat or four-barrel Webers, and power output was listed at 340 horsepower at 7,000 rpm.

The 375 had a robust, ladder-type frame of welded tubes and a 2,600mm wheelbase with 1,325mm front and 1,320mm rear track. The front suspension was unequal length A-arms, Houdaille shocks, transverse leaf springs, and an anti-roll bar. In back was a rigid axle, Houdaille shocks, and semi-elliptic springs. Brakes were finned aluminum drums, and the four-speed transmission was integral with the engine.

Depending upon gearing, top speed nudged 180 miles per hour and had acceleration to match. The model went on a tear, chassis 0360 AM winning the Casablanca 12 Hours in December 1953, and chassis 0370 AM victorious at Buenos Aires in January 1954. In America Jim Kimberly won seventeen of his twenty races with chassis 0364 AM. A 375 MM offshoot (the 375 Plus with a 4.9-liter V-12, de Dion rear, and four-speed transaxle for better weight distribution) won Le Mans (chassis 0396 AM) and the Carrera Panamericana (chassis 0392 AM), ensuring Ferrari's second sports car title in as many years.

Enzo Finds His Tailor

By now World War II was in Europe's rearview mirror, and the Continent was caught up in what became known as the "economic miracle." This decade-plus expansion was brought about by the rebuilding efforts everywhere, and people slowly became comfortable displaying their success.

⌃ Throughout the early 1950s Ferrari used a number of different coachbuilders, but the one he truly wanted was Battista "Pinin" Farina (pictured). "He launched a distinctive style," Ferrari noted in his memoirs, "[and was] the man who transformed the car into a high fashion statement. Pinin was a great artist." *Pininfarina archives*

375 PLUS

Year(s) made: 1954
Total number produced: 8
Hypercars: 1 (0488 AM)
Powertrain: 4,954cc SOHC V-12, 330 hp @ 6,000 rpm; four-speed transmission
Weight: approx. 2,600 lb
Price when new: n/a

PERFORMANCE:
0–60 mph: n/a
0–100 mph: n/a
1/4 mile: n/a
Top speed: estimated 170 mph; (*180 mph seen in the 1954 Carrera Panamericana and "capable of 186 mph flat out in top gear at the Giro di Sicilia")
Road tested by: *both figures from *Ferrari 375 Plus* by Doug Nye
Main competitors: Mercedes-Benz Uhlenhaut Coupe, Pegaso Thrill & Tibidabo, Jaguar D-type

⌃ The dance to get Ferrari and Pininfarina together lasted for the better part of two years. "It was like trying to get two heads of state to sit down at the same table in a neutral country," recalled Sergio Pininfarina, who was one of the men responsible for making it happen. *Pininfarina archives*

"The economy was really growing at that time," remembered Franco Lini, one of Italy's top automotive journalists who would go on to manage the Ferrari competition department in 1967. "There was a real change in the psychology too, for it was several years after the war and people were thinking 'I should buy something good to forget the war.' These were new rich people, and they wanted to demonstrate this by buying a luxury car, or something like that."

Ferrari's success in competition made it the "must have" chassis for the coachbuilders and well heeled alike —and also the most difficult to acquire. According to Giovanni Volpi, Ferrari sales manager Gerolamo Gardini played the market like a fiddle, determining who would get the latest and greatest, who would wait, and who wouldn't get anything at all. Once 375 MM production commenced the clamor began, and all but two chassis would end up with bodies by the carrozzeria more closely associated with Ferrari than any other—Pininfarina.

Company founder Battista "Pinin" Farina (the name became one word in June 1961 on presidential decree) was born in 1893, the tenth of eleven children. From a young age he was self-assured and a quick study, and began working at age eleven in older brother Giovanni's Stablimenti Farina. Farina was a breeding ground for talent, so in 1930 at age thirty-seven, Battista set up his concern, S.A. Carrozzeria Pinin Farina.

By the end of the decade "Pinin's" talent had blossomed, thanks to avant-garde aerodynamic specials like the Lancia Aprilia Berlinetta Aerodinamica. "In the mountains," Pinin noted in his autobiography, "I used to see how the wind sculpted the snow at the edges of the road . . . curved in some places and sharp where it was broken. I wanted to copy those lines for my designs."

After the war a radical Maserati A6 and the landmark Cisitalia 202 berlinetta brought the carrozzeria to the forefront of automotive design, an ascent noticed by Enzo Ferrari. "I was involved in bodywork and had dealings with some of the greatest names in the business," Ferrari wrote. "What I wanted for my cars was character and I found it with the help of Giovanni Battista Pininfarina. . . . It was obvious he was looking for a well-known and beautiful car to dress up, I for a high-class couturier who could adorn my car as it deserved."

According to Pinin's son Sergio, Pinin and Ferrari had a little-known meeting at the 1950 Turin Auto Show. Twenty-three years old at the time, Sergio's childhood had been a masterful indoctrination into the automotive world and its construction techniques. As he grew up the carrozzeria was his second home, its workers an extended family. He obtained a degree in mechanical engineering in 1950 from the Turin Polytechnic and began working full time in the Pininfarina works.

"I was there with them," Sergio recalled of the Turin luncheon. "They only arrived with a 'Pleased to meet you,' and 'I admire very much what you do.' And then there was nothing more."

Thus began an eighteen-month negotiation akin to two heads of state coming together. Pininfarina said Ferrari aid Guglielmo Carraroli tried to break the impasse by issuing an invitation to visit Modena, only to have Pinin reply he'd love to meet, so why didn't Ferrari come to Turin and see his factory? "Both men were prima donnas," Sergio mused. "Mr. Ferrari was *not* coming to Farina, and my father was *not* going to Modena."

A number of months later Ferrari sales manager Gardini suggested they meet halfway in the town of Tortona so Ferrari, Gardini, and the two Pininfarinas gathered for lunch at a restaurant. Once seated, "Everything became extremely easy," Pininfarina recalled. "It was, 'I give you one chassis, and you make one car.' They didn't speak of any price. My father was very enthusiastic; the other man was enthusiastic. . . ."

Pininfarina's first Ferrari appeared in June 1952 (212 Inter cabriolet chassis 0177 EU), and the carrozzeria's work appeared on four other models prior to the competition and

» In late 1953 the Ferrari-Pininfarina relationship really hit its stride with the introduction of the 375 MM and 375 Plus offshoot. This is the Ferrari pits prior to the start of Le Mans in 1954; in the foreground is 375 MM chassis 0380 AM; just behind it is 375 Plus chassis 0392, 375 Plus chassis 0394, and eventual race winner 375 Plus chassis 0396. *The Klementaski Collection*

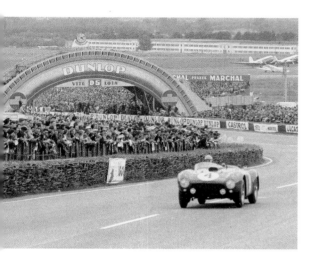

≈ Le Mans is likely the race that cemented Ferrari's global legacy. Here, the Pininfarina-bodied 375 Plus chassis 0396 is on its way to victory in 1954. *The George Phillips collection at the Revs Institute for Automotive Research*

≈ The 375 MM was Ferrari's first hypercar, and the perfect platform for Pininfarina to show its creativity. Chassis 0450 AM was the first of the two 375 MM spyders made for the street and is pictured in Modena's public gardens in 1954. *Pininfarina archives*

"The experience with Ferrari personally was the most exciting and the most difficult. Exciting because it was the most exciting car, difficult because of the highest performance. When you designed a convertible that goes 200 kilometers per hour (125 miles per hour) compared to a convertible that goes 120 kilometers per hour (75 miles per hour), the technical problems multiplied. Ferrari was an experience by itself."

So said Sergio Pininfarina about his early years working with Ferrari, creating what are now some of the most valuable machines in the collector-car kingdom.

The older of Battista and Rosa Farina's two children, Sergio, was born on September 8, 1926, and destined to greatness in the auto industry. When he was a young child, his father demonstrated the basics of proper coachbuilding with a matchbox, first with the center section out, then sliding it back in and encouraging his son to touch it. "See how it is now rigid," his father said. "We want the same when we construct car bodies."

From early childhood Sergio harbored a great drive to excel, completing middle school in seven rather than the normal eight years. He graduated from the Turin Polytechnic in 1950 with a degree in mechanical engineering and marveled at how the mentality of Italy changed in that period. "It was an awakening for a young man like me," he said, showing an appreciation for humanity that never left him. "We had been educated with Fascist ideas and were told in school that the Roman civilization was the most informed. . . . [To] be able to consider more than Rome was an eye-opener, for there was the United States, there was London, France. All these others made great contributions to civilization."

He was soon working full time at the carrozzeria, where he would see how extraordinary his father's "eye" was as they walked the production lines. Pinin would suddenly stop and shout, "There's an imperfection! The right intake is different from the other." A foreman and craftsmen would then come running to make the changes as quickly as possible.

The Ferrari account became his responsibility in 1952 after the luncheon handshake deal between his father and Enzo Ferrari (see page 40). About those trying years: "I felt like I had been thrown in a swimming pool," Sergio summarized. "But my father was confident that I would be able to swim."

That he did, navigating the difficult waters with great equanimity. A side benefit, if one can call it that, was becoming Pinin Farina's development driver. "My father was a pioneer in that he 'sensed' aerodynamics," Pininfarina explained. "When I got my degree, I was an assistant in the aerodynamic branch [so] my father gave me the task to test and develop the Ferraris.

"The aerodynamics had a tremendous importance for noise, rigidity, and the glass. To have a [windscreen] in the right position made the airflow more effortless; to have [it] in the wrong position, the strength of the air was against you—it makes an impression and then a noise. I had to face this program [and] had many, many practical experiences. For example, a windshield wiper was smooth up to 160 kilometers per hour [and] not good over 200. But you had to taste 200 from the wheel, and this was not very easy!

"From the difficulty of testing the car on the road . . . my dream was to make a full-scale wind tunnel. I started to study the idea in 1965, one year before my father died, and finished it in 1970. In 1971 we tested it, and in 1972 it started to operate like an institute, working for Pininfarina."

Sergio fondly (and often) pointed out he could not have accomplished what he did without his brother-in-law, Renzo Carli. "He was a wonderful, very intelligent man, very loyal. He was a very brilliant engineer with a lot of good ideas, more capable of inventing than realizing. He was always inventing something better.

"He was exactly ten years older than I, [and] spent the majority of these ten years in the war—soldier, prisoner of war, so he lost seven years of his life. After the war he became an employee of Shell and wanted to marry my sister. My father met him and said, 'Why don't you come and work with us?' which he did."

On why they had such good chemistry: "I am not a very easy character," Pininfarina said, "and he understood that. He was very happy with his responsibilities and never tried to impose his will on me. We were working together from 1947 up to around the time he died. It was nearly forty years, and it always went well."

About working with "the Sergio" in Modena: "Scaglietti was a very enthusiastic man and a strong character. He was a very good worker with the hands. He experienced very difficult conditions because he was working only with Ferrari, and Ferrari was constantly changing suspension, caster, the inclination, the dimensions of the wheels and tires.

"He was extraordinary at making cars with nothing. He was not extraordinary with the metals and imperfections, but this was impossible [for] he was making cars in ten days. This was the most impossible task given by Ferrari, [who] had a very high esteem, a human esteem, of him."

It was another relationship he cherished: "A funny thing about me and Scaglietti was he was a man with no tradition or culture, but a strength of the nature. . . . When you looked at the shoes, the way he spoke, he was very much a countryman. On the contrary I was young, refined, educated, like a young lord . . . [yet] we were very good friends and had a very nice feeling. All his life, I was 'Engineer Sergio.'"

In fact, the respect between the two was so great that during a prolonged debate over who designed the 250 Spyder California, they signed a document stating the other had created it.

Pininfarina also had the greatest admiration for Ferrari. After being put through the ringer in the early years trying to win over Ferrari, "at my father's funeral," Sergio recalled, "Ferrari unexpectedly said to me, 'from now on, we can use 'tu' with each other." That put Sergio in very rare company, for only Ferrari's closest friends, such as industrialist Pietro Barilla, addressed him in the familiar form.

That close relationship continued another two-plus decades. "In the last years of Ferrari's life," Pininfarina said, "I became like a second son . . . for me he was like a father. . . . Ferrari was a very difficult character [but] in those last years, he was a very sweet man."

⤷ Sergio Pininfarina often rubbed shoulders with some of the world's most famous people—Walt Disney, in this case. Here, he looks younger than the twenty-six he was at the time of the photo. That youthful countenance was a real drawback when he began working with Enzo Ferrari. *Pininfarina archives*

⤷ Sergio Pininfarina gave much credit to his brother-in-law Renzo Carli for helping him build Pininfarina into the coach-building powerhouse it was for a number of decades. Here, the two stand in front of the Pininfarina wind tunnel under construction in 1971. *Pininfarina archives*

» Enzo Ferrari put Sergio Pininfarina through the ringer during their early years of working together, but the two became very close after the death of Battista Pininfarina in 1966. "Ferrari was like a second father to me," Sergio said, and the bond between the two men is evident at the Ferrari annual dinner in 1981.

⌃ The creativity of Battista and Sergio Pininfarina and men such as stylist Franco Martinengo really came to the fore in the summer of 1954 with the sensational one-off 375 MM, chassis 0456 AM. This is inside the Pinin Farina works, shortly before the car made its debut at the Paris Auto Show. *Pininfarina archives*

⌃ Ferrari 375 MM chassis 0456 AM at 1954's Paris Auto Show. A number of design cues were later seen on other Ferraris, and even other makes—the scallop along the side was used for years in Corvettes. *The George Phillips collection at the Revs Institute for Automotive Research*

⌄ A marvelous overview and rare color shot of Ferrari's stand at Paris, where 375 MM chassis 0456 AM was the show's star attraction. The car's first owner was famed film producer Roberto Rossellini. *Gerald Vack photo*

street 375s. While that number of models suggests an easy, smooth relationship, behind the scenes Enzo Ferrari was putting Sergio Pininfarina through the ringer.

"He wanted to work with my father and not me," Pininfarina said, who was put in charge of the account by Pinin on the drive home from Tortona. "I was twenty-five at the time but looked like I was eighteen, just a boy."

Sergio was awed and somewhat intimidated by Ferrari's stature, making the long days in Modena very wearing. He'd awaken at 4:45 to catch the 6:00 a.m. train, and arrive in Modena at 10:00 where someone would drive him to Maranello. But rather than meeting with Enzo, "I was under a certain torture from every manager in Ferrari," he recalled. He endured a relentless pounding, each manager telling him "your quality is not good enough, the weight is too much, and the price is too high, so it's difficult to sell your cars."

The routine would continue for three-plus hours, at which time Ferrari would bring Pininfarina into the office and suggest they go to lunch at Ristorante Cavallino. There, "it was lambrusco, lambrusco, lambrusco, with him drinking one glass while I drank three." After lunch Ferrari excused himself to take a nap, then emerged refreshed late in the afternoon. With young Sergio now more stressed than ever, "That's when we'd discuss prices!" he smiled.

The First Hypercar: The 375 MM

But Pininfarina was a diligent, hard worker and, with his father offering guidance and support, he learned how to effectively (and respectfully) counterpunch. After one particularly arduous trip Sergio returned to Turin and instructed his men "to disassemble the next chassis we get from Maranello so I can know the exact weight of every component." When Ferrari launched into the anticipated tirade on his next visit, Pininfarina calmly reached into his briefcase and produced the itemized list. "Excuse me, Mr. Ferrari," he said, slipping the paper across the desk. "If you look at this you'll see the weight of your components alone, without the body, is greater than the weight you want with our coachwork!"

The complaints stopped, at least for the time being.

"It was lambrusco, lambrusco, lambrusco, with [Ferrari] drinking one glass while I drank three."
—Sergio Pininfarina

⩔ Another one-off 375 MM owned by Rossellini was chassis 0402 AM. It started life as a Pinin Farina spyder; then, after an accident, the car went to Sergio Scaglietti's carrozzeria in Modena for this spectacular coachwork.

Also helping was the MM's success on the track, and it was at this time that Ferrari's first hypercar came into being when several clients ordered examples for the street. The two spyders went to Bao Dai, the deposed emperor of Vietnam who lived in Hong Kong (chassis 0450 AM), and Californian Willametta Keck Day, the daughter of Superior Oil founder William Keck (chassis 0460 AM). Five muscular road-going berlinettas were made for clients in Europe and America; they included Michel Paul-Cavallier, a French industrialist and Ferrari board member (0368 AM); Enrico Wax of Italian spirits importer Vitale & Wax (0378 AM); and northern California advertising executive Alfred Ducato (0472 AM).

Pininfarina also made one of the most spectacular and daring Ferraris ever with chassis 0456 AM. "We were determined to have its fender line an integral design element, something dynamic," remembered Franco Martinengo, another Stablimenti Farina employee who ended up at Pininfarina. "We experimented on scale models and a wooden buck by first trying covered headlights on the sharply creased fenders. We didn't like the way that looked so we only put the turn indicator lights in the fenders."

The end result was a shocker that debuted at Paris in 1954. "One has no hesitation in calling (0456 AM) the most beautiful car of the show," England's *The Motor* effused. Film producer Roberto Rossellini agreed and bought the car.

Amazingly, it wasn't his only one-off 375 MM. Chassis 0402 AM started life as a Pininfarina spyder, but Rossellini had an accident and sent it to Ferrari. "It sat at the factory for a while," Sergio Scaglietti recalled, "but rather than just selling it or getting rid of it, he decided to make it into a coupe."

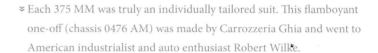
⌃ Ghia's 375 was much more about opulence and luxury than Pinin's creations, as 0476's ornate interior highlights. This one-off has only 14,000 miles on it and, save one repaint to original colors, is in completely untouched condition.

⌃ The simple, functional dashboard and deep bucket seats of 375 MM chassis 0416 AM reflect this Ferrari's quest for ultimate performance above all else. In 1954, this was a legitimate 175-mile-per-hour car.

⌄ Each 375 MM was truly an individually tailored suit. This flamboyant one-off (chassis 0476 AM) was made by Carrozzeria Ghia and went to American industrialist and auto enthusiast Robert Wilke.

≫ Chassis 0488 AM was a one-off 375 MM built for King Leopold of Belgium. It featured a more potent 375 Plus engine, and its design remained a favorite of Battista Pininfarina. *Pininfarina archives*

≫ Here's what made the 375 MM so fast—its formidable *tipo* 102 and 108 4.5-liter V-12s that produced 340 horsepower at 7,000 rpm. In the mid-1950s, a street engine that could turn that high of rpm was *Star Wars*–type technology.

≫ The last 375 MM was chassis 0490 AM that was first seen at the 1955 Turin Auto Show. For decades the car sat in unrestored condition after it had been repainted red. It was then returned to the original color scheme seen here and was a standout on the concours circuit in 2004–05 at venues such as Pebble Beach.

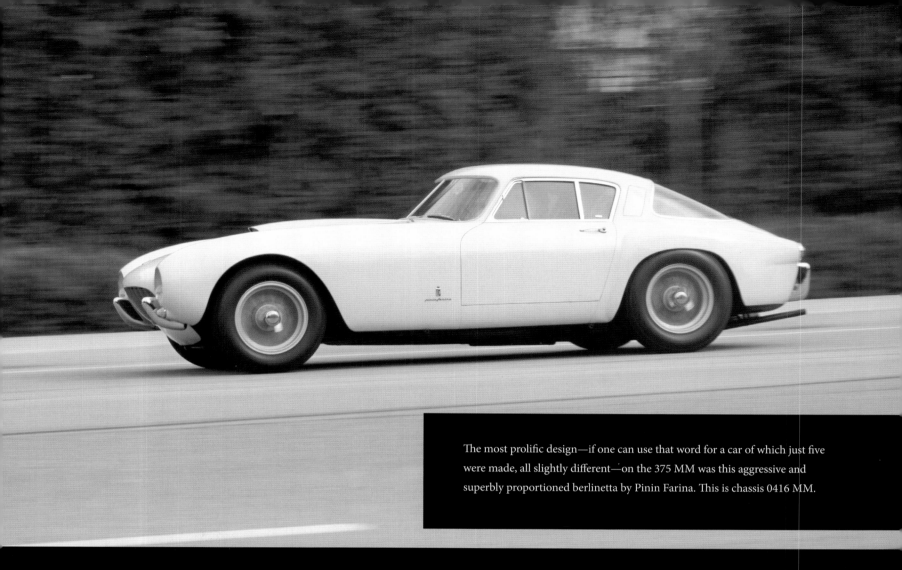

The most prolific design—if one can use that word for a car of which just five were made, all slightly different—on the 375 MM was this aggressive and superbly proportioned berlinetta by Pinin Farina. This is chassis 0416 MM.

What's Behind

THE PININFARINA "LOOK"

There's such a consistency and elegance to Pininfarina's Ferraris, even when a more shocking car like the 375 MM chassis 0456, Modulo, or Enzo was revealed. Why is that?

"There is a definite philosophy of Pininfarina design, one that is not difficult to understand," Sergio Pininfarina explained. "Our mission is to make something new, not to stay where we are—to answer the input we receive from society, from technology, from the changes of life. We want to give new answers in general to the expectations of people.

"At first, it seems to be innovative, but innovation has two strong limitations. One is functionality: You can design a sports car, a family car, anything, but you must always think and be respectful of the function.

"The second is we have some general guidelines that are always the same, from my father down: The simplicity of lines and the

harmony of proportions. When you make a drawing, a project, or a new idea, you keep it simple.

"Simple is much more difficult than making it complicated. Complicated is to impress people, which is something I learned from my father when we were at the London Show in the 1950s. All the journalists were talking of a special Rolls-Royce with gold door handles, and my father was saying, 'All these people are speaking about the handles. It doesn't mean anything. I would like to make a very light handle, out of light alloy.'

"This is the concept of design, as compared to the styling. So I will say again, because it is an important answer. The ideal is innovation, but with two limitations: functionality, and the respect of simplicity, and harmony of proportions. When you make a project, you keep this judgment. It has to have the proportions and keep it innovative. It is a very severe examination."

That's when the humble coachbuilder got the call from sales manager Gardini, who suggested they send it to Scaglietti's Modena-based carrozzeria. Sergio said the design's inspiration came from "following the style of Ferrari. Whether it was Pininfarina or Vignale, I . . . picked up ideas at auto shows and thought, 'I can do that.'" But unlike those two coachbuilders who used sketches, Scaglietti designed by making a thin wire frame over the chassis that showed the outline of the car.

When Sergio's work was completed, he drove 0402 to Modena's Hotel Real Fini, where he found Rossellini in his room with a scantily clad beauty queen. Embarrassed, the bodyman left but was chased down outside, Rossellini beaming with joy upon seeing his car. He then took it out for a test drive, wearing his bathrobe.

Three other one-off MMs were built. Chassis 0476 AM was a two-tone Ghia coupe and found a home in America with Robert C. Wilke, the president of paper products company Leader Cards that sponsored a number of Indy 500 competitors. Chassis 0488 AM was made for Belgium's King Leopold and had a 4.9-liter 375 Plus engine. Pinin thought so highly of 0488's sensational proportions that he dubbed it "the custom car that never ages."

The last 375 MM was chassis 0490 AM. This majestic Pinin Farina berlinetta broke cover at 1955's Turin Auto Show and served as a design forbearer for a number of custom coachwork and competition 250 GTs that followed.

But 0490 was not the era's final race-derived hypercar. In late 1955 Ferrari made four 410 Sports, reportedly to compete in that year's Carrera Panamericana (which was canceled). With a 340 horsepower 4,962cc V-12, five-speed transaxle, and Scaglietti coachwork, the 410 S had brutal acceleration and a top speed second to none—Hans Tanner noted in *Auto Age*'s June 1956 issue that chassis 0598 CM hit 189.72 miles per hour on a long, empty stretch of the autostrada between Rome and Ostia. The second 410 (chassis 0594 CM) was a one-off road-going berlinetta done for the owner of 375 MM 0368, Michel Paul-Cavallier.

410 SPORT

Year(s) made: 1955
Total number produced: 4
Hypercars: 1 (0594 CM)
Drivetrain: 4,962cc SOHC V-12 + five-speed transmission; 380 hp @ 7,000 rpm
Weight: approx. 2,600 lbs
Price when new: n/a

PERFORMANCE:

0–60 mph: n/a
0–100 mph: n/a
1/4 mile: n/a
Top speed: estimated 170+ mph (189.72 mph seen in chassis 0592 CM*)
Road tested by: *Auto Age*, June 1956
Main competitors: Jaguar D-type, Mercedes-Benz 300 SL, and not much else

THE HYPERCAR REFINED: THE 410 SA

By the mid-1950s Italy was transforming itself from a farm-based economy and peasant-culture society into what would become one of the world's top ten economies, as measured by the World Bank and International Monetary Fund. That burgeoning success and evolution into a global powerhouse brought a joy for life not found anywhere else.

"The Italians felt they were free as well as freed from fascism and monarchy, from baronies and agrarian bonds," the University of Palermo's Giovanna Bertelli observed in *Dolce Italia*. "If one were to compare Italy and the republican state to a person's life, one might say these were the youthful years, when anything was possible, when every aspiration could be fulfilled. (Things) forbidden during the twenty years of fascism were now seen as a positive possibility and a promise of new opportunities. Industrialization was the engine not only for the economy but also for social revolution and widespread prosperity."

At the center of Italy's nascent industrial might were its freewheeling *capitani d'industria*. These "captains of industry" "operated in an environment without restraints," said Piero Rivolta, son of one of the Milan area's more noteworthy *capitani* (who would later launch his own line of fast GT cars). "They practically established Italy's postwar industrial style—not only did they find the money, products and people, they had to beat their competition. They also recognized the importance of their employees, so many were like good fathers."

In today's globally connected world with more than 90,000 daily flights, it's easy to forget air travel was spotty several decades ago. Countries such as Italy and America were furiously constructing their highway systems, and the *gran turismo* became the best, most convenient way to travel long distances in comfort and style. As America's *Sports Car Illustrated* observed in the late 1950s, "When a knowledgeable European driver wants to get places in a hurry, he uses a GT-type machine."

Ferrari may have lived to race, but he wasn't blind to what was happening around him. "He was really a marketing genius," Sergio Pininfarina pointed out, noting that Enzo coined terms such as "2+2" and the model name "America." From that genius came his newest car, the 410 Superamerica.

⌃ The other Ferrari hypercar from 1955 with competition roots was 410 Sport chassis 0594 CM. This one-off Scaglietti berlinetta had a 340-horsepower 4,962cc V-12 and, unlike the four-speed 375 MMs, a five-speed transaxle. This was realistically a 180-mile-per-hour machine at a time when the average car struggled to get to half that speed.

In the second half of the 1950s, Ferrari's model lineup began focusing on comfort in addition to speed. The fastest, most expensive offering was the 410 SA that used a V-12 derived from the 410 Sport. This 410 SA, chassis 0479 SA, was owned by espresso machine-maker Renato Bialetti and the Saudi Royal family.

OPPOSITE (TOP):
Superfast I looks sensational next to the River Po in Turin, Italy. The subtle fins were Pinin's homage to America, while the sparse background shows Turin prior to the labor migration boom that would take place in the following decade. *Pininfarina archive*

410 SA

Year(s) made: 1956–1959
Total number produced: 34
Hypercars: Series I, 1 (0483 SA); Series II, 1 (0719 SA); Series III, 12 (all)
Drivetrain: 4,963cc SOHC V12 + four-speed transmission; 380 hp @ 6,500 rpm (0719 SA); 400 hp @ 7,000 rpm (Series III)
Weight: 3,550 lbs
Price when new: $18,500 (0483 SA); $16,000 (0719 SA)

PERFORMANCE:
0–60 mph: "under 5 seconds" (0483 SA); 5.6 seconds (0719 SA); 6.6 seconds (1477 SA, tested at 4,000 ft)
0–100 mph: "11–12 seconds" (0483 SA); 12.1 seconds (0719 SA); 14.5 seconds (1477 SA, tested at 4,000 ft)
1/4 mile: 13.9 seconds @ 108 mph (0719 SA); 14.6 seconds @ 101 mph (1477 SA, tested at 4,000 ft)
Top speed: 161+ mph (0483 SA); 171 mph (0719 SA @ redline in top); 165 mph (1477 SA, tested at 4,000 ft)
Road tested by: 0483, *Road & Track*, July 1957; 0719, *Sports Car Illustrated*, September 1958; 1477 SA, *Road & Track*, December 1962
Main competitors: Maserati 5000 GT, Jaguar XKSS

» The first 410 SA that broached hypercar territory was aptly named "Superfast I." This marvelous Pinin Farina one-off starred at the 1956 Paris Auto Show and used an ultra-potent, competition-derived twin-plug V-12 that allowed the car to break the rear wheels loose when accelerating in third gear. *The George Phillips collection at the Revs Institute for Automotive Research*

Prior to the 410 SA, the common thread among Ferrari's hypercars was all were endurance racers with a modicum of creature comforts. While blisteringly fast, in the uppermost echelons of speed the marketplace was beginning to demand more amenities and refinement. Bentley's 120-mile-per-hour R-Type Continental demonstrated this better than just about anything, as did Mercedes' 140-mile-per-hour 300 SL Gullwing, and even Spain's Pegasos. While most competitors couldn't hold a candle to Ferrari's hypercars in terms of performance, all sold in far greater numbers.

The 410 SA debuted at the 1955 Paris Auto Show as a naked chassis with a 2,800mm wheelbase. Mechanicals followed Ferrari's proven formula with a robust tubular frame and an independent front suspension of wishbones, coil springs, and an anti-roll bar. In back was a rigid axle, semi-elliptic leaf springs, and parallel radius rods. Brakes were large-diameter aluminum drums.

The big news—literally—was the engine. The 4,962cc V-12 was basically the powerplant seen in the 410 Sport racers, with SOHC, a single plug per cylinder, and three 40 DCF Weber carburetors. Power was listed at 340 horsepower at 6,000 rpm.

There would be three different series of 410s, and the first of the 17 Series I cars (chassis 0471 SA) was seen at the 1956 Brussels Auto Show, clothed in sober Pinin Farina coachwork. Next was Ghia's wild one-off "Super Gilda" (0473 SA), done by unheralded design genius Giovanni Savonuzzi (who also did Wilke's 375 MM). Its extravagant coachwork was based on Savonuzzi's fabulous Gilda show car of 1955.

While the "standard" Series I and II 410s were the world's fastest GTs with a top speed of 150-plus miles per hour, most cannot be considered hypercars. The 375 MM performed better by a good margin so the first 410 fitting the "fastest of the fast" definition was one of the decade's most spectacular show cars and carried a moniker worthy of its velocity and appearance. Sergio Pininfarina said the "Super Fast" name came from Ferrari himself, and chassis 0483 SA was the star attraction at the 1956 Paris Auto Show.

≈ Ghia's last Ferrari, 410 SA chassis 0473 SA, certainly looked the part of a hypercar with its wild-winged shape. Yet, neither it nor the traditional Pinin Farina bodied 410s, such as 0479, had enough performance for them to be considered hypercars.

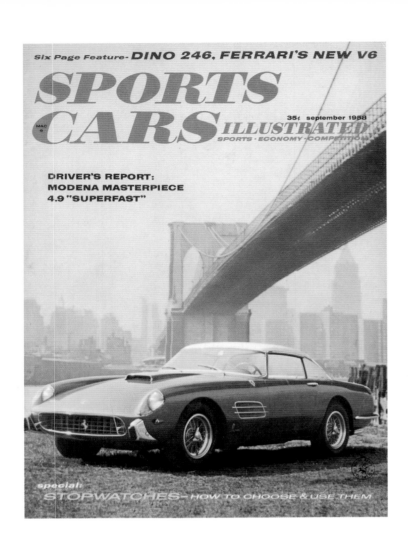

« The second 410 SA hypercar was chassis 0719 SA, also known as "Superfast 4.9." It turned a 13.9 second quarter mile and could clear 170 miles per hour at redline in fourth. "Anybody for pink slips? was *Sports Cars Illustrated*'s summary.

⌄ At the 1958 Paris Auto Show, Ferrari debuted the last of the three series of 410 SAs with chassis 1015 SA. All twelve of these 400-horsepower Series III 410s qualify as hypercars, and they came with both open and covered headlights. *The Edward Eves collection at the Revs Institute for Automotive Research*

« By the late 1950s Ferrari had become a true constructor of road cars with an elevated production line. In 1950 Ferrari produced 26 cars, each made individually. In 1958, the company constructed 183. *The Edward Eves collection at the Revs Institute for Automotive Research*

"We picked up the theme we tried on (375 MM chassis 0456), where we experimented with covered headlights," said designer Franco Martinengo. "Like the earlier car, Superfast I's fender line was something integral and dynamic."

Making the design so fluid was a coach-building trick used by Pinin. "He had his own personal taste," Martinengo noted, "so when a master buck was in front of him he would get a panel beater and place a piece of coachwork over the buck. We usually did this in the sunlight, and would cover the piece with oil to see the reflections . . . to more easily see what we needed to change. We would then put the piece back on the buck, then change and modify the buck. The whole car, all of Superfast 1, was developed using this method."

While Savonuzzi's Super Gilda may have looked like a hypercar, Superfast 1 *was* one. Its wheelbase was shortened by 200mm, and a dual ignition/twin plug competition V-12 was found under its low-slung hood. "A true yardstick of any car's performance is its ability to produce wheelspin on dry concrete road surfaces," *Road & Track*'s July 1957 article noted. "This car can spin its wheels in third gear at 100 mph, and accordingly, acceleration from a standstill to 100 becomes primarily an exercise in the driver's skill." They estimated 0–60 would take "under 5 seconds," 100 in 11 to 12. Top speed was listed at 161 miles per hour, "and possibly more."

The next fire-breathing 410 was "Superfast 4.9." Commissioned by Jan de Vroom, a backer of Luigi Chinetti's North American Racing Team, chassis 0719 SA was the fifth of the six Series II 410s (2,600mm wheelbase vs. the 2,800 for the standard Series I). *Sports Car Illustrated* wrote about its road test in September 1958, listing the engine's output at 380 horsepower at 6,500 rpm, the boost coming from larger Weber carburetors.

The magazine acceleration figures were "the fastest ever recorded [by] SCI," with 0–60 taking 5.6 seconds, 100 12.1, and the quarter mile 13.9. If 0719 had been run to

⩦ An example of an open-headlight Series III 410 was chassis 1477 SA, owned by hotel magnate Bill Harrah. *Road & Track* commented the car's "forward thrust has to be felt to be believed."

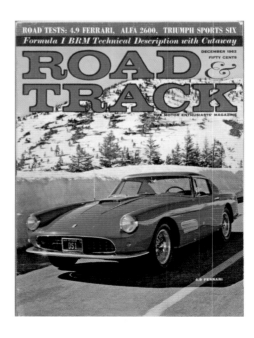

redline in fourth gear, top speed would have been 171 miles per hour. "Anybody for pink slips?" was the test's most telling line.

A little more than a year later, the first Series III 410 (chassis 1015 SA) debuted at the 1958 Paris Show. All twelve of these Ferraris had supremely elegant Pinin Farina coachwork and qualify as hypercars. The 4,962cc V-12 was massively re-engineered with outside plug heads (the first for a road-going Ferrari), new connecting rods and a new crankshaft, and made 400 horsepower. Brake size increased, and a new four-speed gearbox was used.

Road & Track tested chassis 1477 SA, hotel magnate Bill Harrah's car. Even though they experienced "considerable clutch slippage" from its nonstock unit while testing at a power-sapping 4,000 feet, it still hit 60 in 6.6 seconds, 100 in 14.5, and topped out at 165 miles per hour. "There's no sound like the scream of a twelve-cylinder Ferrari engine," the magazine observed, "and the forward thrust has to be felt to be believed."

THE HEIR APPARENT

At the 1960 Turin Auto Show, Ferrari seemed destined to keep the "world's fastest" title when Pininfarina displayed a slippery-looking shape called "Superfast II." Designed by Aldo Brovarone, "I always liked the Vanwalls that raced in Formula 1," he recalled, "and the design of those cars inspired me. In the beginning the nose was too pointed, too thin, so we opened it upward to make it look more like a Ferrari. I also looked at our Abarth Record car and its themes for aerodynamics."

Built on 400 Superamerica chassis 2207 SA (the model was introduced in late 1959), the prestigious annual *Automobile Year* noted Superfast II "aroused so much interest that [Pinin] had to build five more to order. . . . Speeds up to 180 mile per hour are claimed." The clamor was actually much stronger—nearly 75 percent of the 400 SAs built had Superfast II-based *coupe aerodinamico* coachwork.

But not one was a hypercar. Looking at that clutter-free form, 180 miles per hour seemed believable, but the only 400 that approached that number was chassis 2861 SA. Originally shown at the 1961 Paris Auto Show, Bill Harrah bought it and had ace mechanic Bill Rudd modify the four-liter engine to produce 400 horsepower, according to *Road & Track*'s 1963 test. However, acceleration times—60 in 7.8 seconds, 100 miles per hour in 18.7—were anything but hypercar quick. Indeed, the 250 GT *R&T* tested in 1960 was faster to 60, 100, and through the quarter mile.

It mattered not. While Maserati's limited-production 171-mile-per-hour 5000 GT was the hypercar to beat in the early 1960s, the 400 SA outsold it with forty-seven made between 1959 and 1964.

Plus Enzo and his men weren't sitting still. Ferrari's next hypercar would be an entirely different animal, one capable of leaving the 5000 GT—and all others—in its wake. Then his hypercars would undergo a real revolution by putting "the ox behind the cart," rather than leaving it in front.

≫ This dramatically shaped Ferrari wowed everyone at 1960's Turin Motor Show and was called "Superfast II." Underneath that avant-garde shape was a 400 SA platform (chassis 2207 SA) with a 340-horsepower V-12. While top speed was quoted at 180 miles per hour, this was very likely overstating the case. *LAT*

3

MOVE TO THE MIDDLE

(1963–1983)

photo from March 3, 1963. The new mid-engine 250 P (left) makes an interesting contrast to the 330 LMB, one of Maranello's last front-engine warriors; it would be several years before the mid-engine revolution hit Maranello's road cars. Drivers are Mike Parkes (in the 250 P), with (left to right) Ludovico Scarfiotti, Lorenzo Bandini, and Nino Vaccarella behind him. Willy Mairesse is in the LMB.
Franco Villani/ ATS Photo Library

⌃ A close relative of the 250 GTO was the 330 LMB. Just four were built in 1963, and one (chassis 4619 SA) was used during the Le Mans test days where it hit 186 mph. The car was then was sold to Luigi Chinetti who used it on the street. Pictured is a sister car (chassis 4453 SA) that was also sold through Chinett.

⌄ In the late 1950s endurance racing limited engine sizes to 3 liters for championship contending cars. The ultimate expression of such machines was Ferrari's immortal 250 GTO. Though the model won the title from 1962 to 1964, and some were driven on the street, it was not a hypercar because it focused on all around balance, rather than out and out top speed.

330 LMB

Year(s) made: 1963
Total number produced: 4
Hypercars: 2 (4619 SA; 4381 SA [transformed into road barchetta in 1964 by Fantuzzi])
Drivetrain: 3,967cc SOHC V12, 390 hp @ 7,500 rpm; four-speed manual transmission
Weight: 2,094 lb (competition guise)
Price when new: n/a

PERFORMANCE:
  0–60 mph: n/a
  0–100 mph: n/a
  1/4 mile: n/a
  Top speed: 177 mph
Road tested in: *Car & Driver*, November 1963 (4619 SA)
Main competitors: Ferrari 250 P, 250 GTO, Ford GT40, Maserati Tipo 151

In 1964, Carrozzeria Fantuzzi in Modena transformed 330 LMB chassis 4381 into a rakish barchetta via this startling rebody. *Marcel Massini archive*

AT FERRARI'S PRESS CONFERENCE IN FEBRUARY 1962, Enzo introduced one of the greatest cars ever, the 250 GTO. This spectacular Scaglietti-designed, 300-horsepower berlinetta would go on to win the next three sports-racing championships, but that didn't prevent Ferrari and his men from trying to make a world beater go faster. So once again they resorted to shoehorning a bigger engine between the front fenders.

These 250 GTO "cousins" carried the model designation "330." Power came from a modified version of the 400 SA's 3,967cc V-12. With more radical cams, higher compression, and more carburetion, power jumped to 390 horsepower at 7,500 rpm. The engine also had considerably more torque than the 250 GTO's 2,953cc twelve, so some of the 330s used four-speed gearboxes.

There were two body styles. The 330 GTO was nearly identical to the 250, and three were produced in 1962–63. The 330 LMB was designed by Pininfarina and looked like the carrozzeria mated the front of the 250/330 GTO with the rear of Ferrari's newest production model, the 250 GT/L or "Lusso."

Four LMBs were made in 1963, and one became Ferrari's (and likely the world's) fastest-ever road car. Chassis 4619 SA served as the development mule, and test driver Mike Parkes sorted it and the other LMBs around the Modena area and on the autodromo track. As Graham Gauld recalled of one memorable drive with Parkes in *Modena Racing Memories*, "[He] had the look of a slightly fey English schoolboy but put him behind the wheel of a racing car and he was transformed into a racing animal."

That "animal" drove 4619 during the Le Mans test days on April 6–7. According to authors Keith Bluemel and Jess Pourret in *Ferrari 250 GTO*, it became the first car to reach the magic 300 kilometers per hour (186 miles per hour) down the famed Mulsanne Straight. That was the closest the 4619 ever came to racing, for it was then sold off with a fully upholstered interior and remained a road car its entire life.

(As a side note, another 330 LMB was transformed into a hypercar when Modena's Carrozzeria Fantuzzi rebodied 4381 SA in 1964 for Rome-based production company Cromograph Srl. Four years later, the Ferrari and its rakish barchetta coachwork was featured in "Toby Dammit," the last segment of the omnibus film *Histories extraordinaires*.)

THE "OX AND CART" CHANGE POSITIONS

Though the 250 GTO won the championship from 1962 to 1964, it and the 330 LMB were a dying breed. Their demise unknowingly began on January 19, 1958, in Buenos Aires when Ferrari's 2,417cc, 280-horsepower 246 F1 car was thoroughly trounced by a 1,960cc 170-horsepower Cooper-Climax. The English car was an unknown to Ferrari, its mid-engine design a first for postwar F1.

That surprising victory startled and intrigued chief engineer Carlo Chiti. Was the "ugly little brute" (as he would call the Cooper) responsible for the win, or was it the talent of Stirling Moss, one of the greatest drivers ever? His answer came at the next race in Monaco when Maurice Trintignant, who had only one win in his up-to-then 46 F1 starts, took the checkered flag.

While Ferrari would go on and clinch the 1958 F1 title, "the Coopers' results made us think their secret [was] in their exceptional lightness," Chiti told author Piero Casucci in *Chiti Grand Prix*. "We did not give much consideration to their road holding" until testing convinced Chiti the mid-engine design was indeed superior.

The configuration proved to be a hard sell in Maranello. Engineer Gioachino Colombo had discussions about a rear engine car with Enzo before World War II, to which Ferrari would reply, "It's always been the ox that pulls the cart."

Chiti encountered the same resistance. "One of the stumbling blocks was Ferrari built and sold GTs with a front engine," the engineer recalled. "[If] we went ahead with and opted for a rear engine for the racing cars, it was feared that this would disorient the public. I thought this was unfounded: What really counted was winning races."

Cooper helped Chiti's cause when it won the 1959 and 1960 drivers' and constructors' titles. On May 22 ,1960, Ferrari tested its first mid-engine F1 car (the 246); a week later in Monaco it placed sixth. Development continued apace, and in 1961 Ferrari's mid-engine 156 F1 won the championship.

≋ In 1961, Ferrari fielded its first championship winning mid-engine car, the 156 F1. This marvelous, behind-the-scenes look at testing shows the 156 on stands in the pits at Modena's Aeroautodromo. In front of the car is chief engineer Carlo Chiti, while Enzo Ferrari sits on the pit lane wall. *Peter Coltrin at the Klementaski Collection*

« And here is the result of testing: Ferrari 156 F1s at Dutch Grand Prix in 1961. In front in car No. 1 is race winner Wolfgang Von Trips, and right behind and finishing in second place is Phil Hill. Hill would win that year's driver's championship. *The George Phillips collection at the Revs Institute for Automotive Research*

« Engineer Carlo Chiti also wanted the endurance racers to use the mid-engine configuration. In 1961 he got his wish, though the first car to use it was 246 SP chassis 0790, seen at its debut at Sebring, where it was a DNF. *The George Phillips collection at the Revs Institute for Automotive Research*

« Upheaval hit Maranello in late 1961, when Enzo Ferrari fired Chiti and other key administrative and engineering personnel. The calamitous event thrust twenty-something and very green Mauro Forghieri into the spotlight. "I was one of the few engineers remaining," he remembered of the surprise promotion. "The Old Man offered to me to take care of the racing, and made it very clear he was behind me 100 percent."

⌃ Forghieri finally succeeded in getting a mid-engine V-12 endurance racer in 1963. The result was the overall win at Le Mans in 1963 with 250 P chassis 0814 (pictured right after the victory). *The George Phillips collection at the Revs Institute for Automotive Research*

⌄ Several hearty clients ordered 250 LMs for use on the street. One such individual was southern Californian Steve Earle, who took this photo of chassis 6107 GT on Mulholland Drive shortly after taking delivery in late 1964. Earle later became known as the father of the Monterey Historic Races. *Steve Earle*

⌃ The Pininfarina-designed 250 LM was basically a 250 P with a roof. The 180-mile-per-hour model would be a robust warrior for Ferrari on the racetrack, and here is the prototype (chassis 5149 GT) at the 1963 London Motor Show. Thirty-two 250 LMs would be built over the next two years. *LAT*

> " **Everyone at Ferrari was certain a twelve-cylinder rear-engined sports racing car would be technical madness.** "
>
> — **Mauro Forghieri**

Several weeks after clinching both the F1 and endurance-racing crowns, Maranello was rocked when eight key personnel—sales manager Gardini, engineers Chiti and Giotto Bizzarrini, racing director Romolo Tavoni, and four others—left Ferrari. Chiti, Bizzarrini, and Tavoni all told the author the mass exodus was caused by Ferrari's firing of Gardini after he confronted Enzo about wife Laura's continual meddling in company affairs. They banded together to try to get Ferrari to bring Gardini back, and Enzo promptly fired them. The group would soon form the fascinating but short-lived Ferrari competitor, ATS.

With his top engineering talent gone, Ferrari turned to Mauro Forghieri, a young engineer not even twenty-six years old at the time. "I couldn't have imagined such a thing," Forghieri reflected. "They were very, very good people, but Enzo was a very strong man, and I know that he would not accept management criticizing [him] that way."

With Ferrari telling Forghieri that he would take care of the politics so the engineer could focus on the cars, "In reality, the chassis, suspension, and rigidity were our weak points," Forghieri wrote in his fascinating book, *Forghieri on Ferrari*. "It was the advent of the rear engine that caused a large number of new problems, because front engine cars were far less demanding on the road holding front. . . .

"Everyone at Ferrari was certain a twelve-cylinder rear-engined sports racing car would be technical madness. 'It can't work with all the weight in the back,' the Commendatore maintained, as he was one of the most resistant to changing ideas. However, I was sure it would be successful and proposed we build a car with a rear end weight no more than 55 percent of the total. . . ."

The result was the 250 P, and in 1963 that Ferrari became the first mid-engine car to win Le Mans (chassis 0814). The 250 P also served as the basis for the 250 LM, an endurance racer that would become the world's first mid-engine hypercar.

The LM debuted at the 1963 Paris Auto Show (chassis 5149 GT), carrying a rakish design by Pininfarina that was, in essence, a closed version of the 250 P with a roofline and rear end treatment nearly identical to Pininfarina's 250 GTO/64. The LM shared the P's 2,400mm wheelbase, 1,350mm front track, and 1,340mm rear track. Both had tubular space frame chassis and independent front and rear suspensions of double wishbones, coil springs over telescopic shocks, and anti-roll bars. Brakes were large Dunlop discs, inboard in the rear, and the rear-mounted gearbox was a five-speed.

Only the Paris Show Car would have a 2,953cc V-12; the remaining thirty-one were fitted with 3,285cc V-12s with SOHC, single-plug per cylinder, a 9.7:1 compression ratio, and six 38 DCN Weber carburetors. Power output was 320 horsepower at 7,500 rpm. In April 1964, Ferrari asked the CSI to have the car homologated to run in the GT class. The petition was denied, forcing the LM to run as a prototype against his own, faster cars, and Ford's new GT40.

The LM was produced in 1964 and 1965, and it was sold to a number of Ferrari's top customers, including Chinetti's NART team, Maranello Concessionaires of England, and Jacques Swaters' Ecurie Francorchamps of Belgium. The model proved quite successful, scoring a number of wins in 1965. Indeed, Chinetti's overall victory at Le Mans in chassis 5893 GT remains Ferrari's last at the event.

≫ In 1965 Pininfarina attempted to civilize the LM with a spectacular one-off design exercise on chassis 6025 GT. Designer Leonardo Fioravanti said he redid most every panel in an effort to make the car more aerodynamic, the most noticeable difference being the sloped rear glass.

« The fully upholstered interior showed Pininfarina's attention to making the car more useable on a regular basis. The red, white, and blue color scheme of the interior and exterior was a subtle nod to the car's first owner, American importer Luigi Chinetti.

The LM was conceived for racing, yet that didn't stop six of the first owners from using them as road-going hypercars (chassis 5903 GT, 6025 GT, 6045 GT, 6107 GT, 6217 GT, and 6233 GT, according to research by historian Marcel Massini). Five were "normal" LMs, while one (6025 GT) was an extraordinary one-off done by Pininfarina.

Penning the redesign was the company's incredibly gifted young stylist, Leonardo Fioravanti. Ambitious, cultured, and elegant, he was just twenty-six at the time and recalled that "it was my first piece of work at Pininfarina. My goal was to make the shape as aerodynamic as possible." He accomplished this by redesigning every body panel and penning a lovely, sloping rear windscreen that ended near a subtle swept-up rear spoiler.

Chassis 6025 was one of the stars at the 1965 Geneva Auto Show, shown with white paint, a blue stripe down the center, red leather upholstery, and a top speed that was later claimed to be in the region of 180 miles per hour. "There was much interest in the only news from Pininfarina: The Ferrari 250 berlinetta Le Mans," Italy's *Quattroruote* noted. "The lines are aggressive and very pleasant . . . with a well-shaped back. It is a unique example and will be next exhibited at the New York Auto Show." There it appeared on Chinetti's stand, and Luigi subsequently owned it for a number of years.

≈ When viewing the one-off 250 LM chassis 6025 GT from the side, one can more easily see the differences between it and the standard LM; this includes a more sleek nose, redesigned rear fenders, and the lovely slope of the unique rear glass.

PAY ATTENTION TO THE HAT

Enzo Ferrari was a master motivator who did whatever was necessary to achieve the results he desired. Carrozzeria Touring design head Carlo Anderloni, a very even-tempered and friendly man, got an unexpected taste of Ferrari's volatile side in 1949 while his craftsmen toiled away on 166s for the Mille Miglia.

Anderloni recalled Ferrari visiting the Touring works in Milan to examine the barchettas undergoing construction. But rather than having a cordial visit, shortly after Enzo arrived he launched into a tirade, screaming the work was unsatisfactory, moving too slowly, and that he might cancel his order. He then stormed out and drove back to Modena.

Stunned, Anderloni looked over at business partner Gaetano Ponzoni, and the two quickly retreated to the offices. They determined Carlo would drive down to Modena the following day and see Ferrari to smooth over the waters. Because Italy's postwar

infrastructure was still in disarray, the trip was arduous and Carlo arrived at the factory late in the morning. He was immediately escorted into Enzo's office and invited to sit down.

"What brings you here?" Anderloni remembered Ferrari saying. "This is a wonderful surprise."

"Yesterday's blow up has us a bit distressed," Carlo replied. "What can we do to help the situation?"

"Yesterday?" Ferrari responded, dismissing the episode with a shrug of the shoulders. "Shall we go get an early lunch?"

Sergio Pininfarina saw much of the same with people in Maranello and offered an explanation: "I think Ferrari was hard on the racing drivers because he was concerned about the mechanical parts of his car more than he was about the performance of the drivers; the human side came in second. Therefore, if a driver dared to say, 'I lost because a Ferrari was not good . . .'"

Pininfarina then let out a gasp and said, "Can you imagine?

"Back then he was used to winning almost every race, and when I was in Maranello on Monday (after a victory), he was terrible, in very bad temper. You know why? Because he was afraid his people were sleeping, that they would relax. So he was very difficult, very, very hard—in the words, the attitude."

Pininfarina then recalled the other side of the man: "In the few cases when Ferrari was beaten and there was a defeat, I saw him very encouraging, in good temper. He understood that if he was [hard], he would destroy the morale of his army. He was a very intelligent man."

For Sergio Scaglietti, the key to gauging Ferrari's mood was how he wore his hat. If he was cajoling, friendly, the hat was on straight. But once it tipped to a side, look out! There was an explosion coming.

Unlike Carrozzerias Pininfarina and Touring, which made sketches prior to cars being built, Sergio would lay a wire frame over the chassis to show what the body would look like. He said Ferrari normally left him alone during the process, two exceptions being the times the hat went crooked while he toiled away on the Daytona prototype. "He came here [to the carrozzeria] and didn't [like] the job we did," Scaglietti explained. "He didn't like the front or the back and went crazy. So we cut the front and replaced it with something we would do, like the 375 MM [done for Rossellini].

"The other time was when the Ferrari technical team felt the car was too narrow, so we went in and cut it right down the middle, adding 4 to 5 centimeters more. That's how the Daytona was born."

For which enthusiasts can thank a crooked hat.

THE GAME CHANGES—IN MORE WAYS THAN ONE

When the 250 LM was introduced, Sergio Pininfarina and brother-in-law Renzo Carli had been running Pininfarina's operations for several years. During this period Sergio's working relationship with Enzo Ferrari improved dramatically; one episode from the mid-1950s signified the thaw.

After Pininfarina had endured another punishing round with Ferrari's purchasing managers, a well-rested Enzo called him into his office. "Every day I come here," a worn-out Sergio said, taking a seat, "I become ten years older! I wake up early in the morning, and then come here. Your men hit me over the head for three-plus hours. Then we go to lunch and drink and drink. Then you take a nap. . . ."

Pininfarina stopped midsentence and looked up. Ferrari was laughing.

≈ Sergio Pininfarina's joy of seeing his Dino show car unveiled (seen below) was short-lived. One month later, at the 1965 Turin Auto Show, Lamborghini unveiled this radical P400 chassis, with a sidewinder V-12. After wandering over to take a look, "I was envious when I saw that chassis," he recalled. *LAT*

250 LM

Year(s) made: 1963–1965
Total number produced: 32
Hypercars: 6 (5903 GT, 6025 GT, 6045 GT, 6107 GT, 6217 GT, 6233 GT
Drivetrain: 3,286cc SOHC V-12 + five-speed; 300–320 hp @ 7,500 rpm
Weight: 1,800+ lb*
Price when new: $15,600 (standard LM); $20,000 (U.S. price)**

PERFORMANCE:
0–60 mph: n/a
0–100 mph: n/a
1/4 mile: n/a
Top speed (mfr): up to 180 mph, depending upon final drive ratio
Road tested by: *Motor Racing*, April 1965; **Car & Driver*, May 1965
Main competitors: Ford GT40; Shelby Cobra Daytona Coupe

≈ Perhaps the biggest proponent for a proper mid-engine Ferrari road car was Sergio Pininfarina. After pushing on Enzo Ferrari for quite some time he got his wish, the result being the one-off V-6-powered Dino Speciale (chassis 0840) that debuted at the 1965 Paris Auto Show. It was the first step to the production 206/246 Dino that would appear in June 1968. *Pininfarina archives*

⌃ Lamborghini's P400 chassis was the basis for the Miura that debuted six months later at the 1966 Geneva Auto Show. With its sensational looks and avant-garde mechanicals, this was the car that made ultimate performance go mainstream. "Every rich and impatient man wanted one," remembered the car's father, Gianpaolo Dallara. *LAT*

« Being the world's fastest road car wasn't the only place where Ferrari was under assault. In 1963, the Ford Motor Company's attempt to buy Ferrari was rebuked at the last moment. "That's okay," Henry Ford said. "Let's go beat them." Ford's weapon was the GT40; here, chassis GT/101 is being assembled in Ford Advanced Vehicles on March 28, 1964. *The George Phillips collection at the Revs Institute for Automotive Research*

"It was a long struggle," Sergio reflected, "but I think Mr. Ferrari began to appreciate my sense of goodwill and commitment to the work. I was a very serious young man, very fastidious, a committed, precise engineer, so he started to say, 'He is reliable.' Then he started to say, 'He is a good engineer.'"

As the trust built, one topic on which they couldn't agree was that old sticking point of Enzo's—the placement of a car's engine. "I was insisting that we should make a mid-engined car," Pininfarina noted. "Mr. Ferrari said, 'For racing, yes, but it's too dangerous for the customers.' So I insisted, and insisted, and all the salesmen were with me.

"When he finally agreed, he said, 'Okay, you make it not with a Ferrari (badge), but with a Dino.' This was because the Dino was a less powerful car, and in his idea, less powerful meant less danger for the customers."

Pininfarina's landmark prototype debuted at the 1965 Paris Auto Show with curvaceous coachwork by stylist Aldo Brovanone and a longitudinal 1986cc V-6 behind the driver. The Dino was splashed across the pages of automotive magazines worldwide, some noting Ferrari's success with the Dino 166 P endurance and hill-climb racer, upon which the mechanicals were based.

Then everyone's attention shifted. A month later at the Turin Show Lamborghini displayed its naked "P400" chassis. The advanced design was the work of chief engineer (and former Ferrari employee) Gianpaolo Dallara and had a deep section steel platform with independent suspension front and rear and large Girling discs. The 3,929cc DOHC V-12 was cleverly in unit with the gearbox and placed sideways just forward the rear wheels.

"The prize for the big poker hand was reserved for Comm. Lamborghini (who) showed up with . . . a wild new transverse-engined chassis," *Road & Track*'s Turin show coverage noted. "Various Ferrari bods were seen drifting by in overcoats, with collars turned up to have a look at it."

One was Sergio Pininfarina. "When I saw that chassis," he recalled, "I was envious." He felt much the same when the stunning Bertone-bodied Miura appeared six months later at the Geneva Show.

THE MOST EXOTIC OF ALL: THE 365 P

Pininfarina would get his chance to counter the game-changing Lamborghini, but it would come after creating a Ferrari that would leave the Miura in its dust.

This hypercar's birth can be traced back to early 1963 when Henry Ford and Lee Iacocca were interested in international racing, and figured buying Ferrari was the fastest way to gain expertise. America's youth movement was right on the horizon, and the *La Dolce Vita* lifestyle had spread out beyond Rome to the French and Italian Riviera and elsewhere. Simply put, performance and style were in, and Ford knew it.

So the company pursued Ferrari. According to *Road & Track*'s October 1966 article on the subject, Ford's pointman on the task VP and general manager Donald Frey intoned that after proper due diligence had been done, the actual purchase negotiations were serious and intense but collapsed after ten days. Frey returned to America disappointed but not surprised, Henry Ford famously saying at the end of the briefing, "That's okay. Let's go beat them."

In *Forghieri on Ferrari*, the engineer said the negotiations "started because of the financial difficulties Ferrari faced. . . . The brand was established and the number of

> " **I was insisting we should make a mid-engine car. Mr. Ferrari said, 'For racing, yes, but it is too dangerous for customers.'"**
>
> **— Sergio Pininfarina**

Ferrari handily beat Ford at their Le Mans showdown in 1964. Then the rivalry intensified considerably, and a number of voluptuous prototype racers came out of Maranello in the mid-1960s. This included the 365 P2 chassis 0828 at Daytona in 1966. Its 365 P sister model would act as the underpinnings for an extraordinary hypercar a few months later. *The Stanley Rosenthal collection at the Revs Institute for Automotive Research*

OPPOSITE: According to Aldo Brovarone, the designer of the 365 P, the three-abreast seating was done so the Ferrari could more easily accommodate its commissioning client, Fiat's Gianni Agnelli (second from left, looking into car). With him are Ferrari's Mike Parkes (left), Sergio Pininfarina (behind Agnelli), and Renzo Carli (far right). *Pininfarina archives*

365 P

Year(s) made: 1965–1966
Total number produced: 3
Hypercars: 2 (8815, 8971)
Drivetrain: 4,390cc SOHC V-12 + five-speed transmission; 380 hp @ 7,300 rpm
Weight: n/a
Price when new: n/a

PERFORMANCE:
  0–60 mph: n/a
  0–100 mph: n/a
  1/4 mile: n/a
  Top speed (mfr): listed at 186 mph
Road tested by: Leonardo Fioravanti and the owners!
Main competitors: Ford GT40; an airplane

road cars produced each year increased, but that wasn't enough to pay for Maranello's involvement in the Formula 1 World Championship and sports cars racing."

When the deal fell through and Ford made its racing ambitions clear, Forghieri writes "despite their extensive means, Ford didn't especially impress us [at Le Mans in 1964]. The following year, though, we began a period of work that was grueling. The Ford challenge had become incandescent [and] I talked to Enzo Ferrari about the situation. . . . His pride became stimulated by challenges, particularly those that were weighted against us."

Indeed they were. The purchase of Ferrari would have amounted to only a rounding error for Ford; now, it was using that cubic checkbook to best Maranello at the world's greatest, most prestigious race. With Enzo steering the ship through combat, the young engineer was thankful to be surrounded by "the incredible ability of all the mechanics" and "people who could solve any problem."

Over the next four years, Ferrari produced a dizzying array of prototype racers. In 1964, the 250 P morphed into the 275 P and 330 P (the former winning that year's Le Mans with chassis 0816), which then begat the 275 P2 and 330 P2 in 1965. The latter showed considerable advancement over its forerunners—changes included riveting the aluminum body panels to the chassis for additional rigidity, suspension that used F1 practices, and a new 3,967cc DOHC V-12.

The 365 P was born after the P2s and was essentially the 330 fitted with a more powerful (and torquey) 4,390cc SOHC V-12. Just one competed in the 1965 season (chassis 0824), and the model served as the basis for the most exotic hypercar yet seen, Pininfarina's 365 P of 1966.

With a quoted (and probably realistic, given the car's competition underpinnings) top speed of 300 kilometers per hour (186 miles per hour), the first 365 P appeared at the Paris Auto Show (chassis 8971), and was sold to Luigi Chinetti. The second (chassis 8815) was shown a few weeks later at Turin and went to Fiat patriarch, Gianni Agnelli. Both featured central steering with three-abreast seating (decades ahead of McLaren's F1) and a large glass roof. Agnelli's car had a rear spoiler that gave it a more aggressive look.

Designing the 365 P was Aldo Brovarone. "It was done for Agnelli," he recalled. "I had a completely free hand in designing the car, and the inspiration came from the Berlinetta Dino so I made the shape larger and used the front of the 500 Superfast [Ferrari's luxurious limited production GT that succeeded the 400 SA]."

About the central steering: "That was done to make it easier for Agnelli to enter and exit," Brovarone said. "He had a very stiff leg from an automobile accident, and the

⊬ Several months after the Miura's 1966 Geneva Show debut, Ferrari and Pininfarina teamed up to create what was likely the most exotic car made to date. The mechanicals were based on Ferrari's 365 P endurance racer, and the car featured central steering with three abreast seating. This Ferrari (chassis 8815) is the second of the two cars. *Pininfarina archive*

⋩ The first of the two 365 Ps was chassis 8971; it debuted at 1966's Paris Auto Show (pictured). The car was made for American importer Luigi Chinetti, and is still owned by the Chinetti family. Designer Aldo Brovarone said the design's inspiration came from the Dino and limited production 500 Superfast. *LAT*

Though Ferrari and Pininfarina displayed several mid-engine show cars in the mid-1960s, Maranello's best all-around performing production model remained the front-engine 275 GTB, seen here with Battista Pininfarina before he passed away in 1966. *Pininfarina archives*

rotating seat and central position gave space for him to slip his leg into position under the steering wheel. Agnelli's car had a spoiler mounted on the rear because of tests that were carried out by Ing. Fioravanti, who noticed instability at high speeds."

While the press praised the 365 P, the period was anything but easy for Sergio Pininfarina. Several months before the P's unveiling "my father died, and he was very popular, very much loved by the people. The sympathy that people had transferred from my father to me, and I felt a lot of warmth around during the 1960s."

One person offering heartfelt condolences was Enzo Ferrari. "At my father's funeral," Pininfarina continued, "he said, 'From now on we can use *tu*,' the familiar form of addressing someone. That meant a lot to me, for it expressed his trust."

ULTIMATE PERFORMANCE GOES MAINSTREAM: THE HYPERCAR NEEDLE DOESN'T MOVE

As Ford's GT40 staged its assault on Ferrari's competition P cars, a number of competitors were doing the same to Enzo's street offerings. In 1967, Ferrari's fastest production car was the 160-mile-per-hour 275 GTB/4; five years earlier, that speed was nudging hypercar territory.

With the *La Dolce Vita* lifestyle in full swing, "whether you went from Milan to Turin in thirty-one or thirty-two minutes became very important, something that was spoken about at cocktail parties," said Piero Rivolta, then CEO of Ferrari competitor Iso. He would know, for his 365-horsepower Grifo could clear 160 miles per hour. Maserati's Ghibli had just gone on sale and had a claimed (but unrealistic) top speed of 174. Even small manufacturer Bizzarrini's GT Strada would touch 160 when optioned the right way.

≈ In the second half of 1966, Ferrari installed a high-revving 300-horsepower four-cam engine (shown) in the car; that model was known as the 275 GTB/4. *The Karl Ludvigsen collection at the Revs Institute for Automotive Research*

≈ Over the course of the 1960s, Ferrari transformed itself from an artisan manufacturer into a true industrial concern. Production would top 900 in 1970, an 800 percent increase in thirteen years. Along with that came enormous labor issues, Sergio Scaglietti saying strikes and unrest instigated Ferrari to sell 40 percent of his company to Fiat in 1969. *John Clinard*

Ferrari's on-road performance supremacy was under threat as never before in the late 1960s. The world was caught up in the allure of glamor and speed, and cars such as a properly optioned Maserati Ghibli (pictured) could touch 160 miles per hour, while the Miura was exceeding 170 in road tests.

≈ Ferrari formulated a response to its contenders throughout 1967: The model's designation would be the 365 GTB/4, better known as the Daytona. This one-off (chassis 11001) was the second Daytona prototype, and was made in early 1968. With the Daytona, Enzo preferred to "keep the ox in front of the cart," rather than going the mid-engine route.

365 GTB/4 & GTS/4 (DAYTONA COUPE AND SPYDER)

Year(s) made: 1968–1973 (coupe); 1969–1973 (spyder)
Total number produced: 1,406 (1,284 coupes, 122 spyders
Hypercars: all
Drivetrain: 4,390cc DOHC V-12 + five-speed transmission; 352 hp @ 7,500 rpm
Weight: 3,377 lbs
Price when new: $19,500

PERFORMANCE:

0–60 mph: 5.8 seconds
0–100 mph: 12.8 seconds
1/4 mile: 13.2 seconds @ 110 mph*
Top speed: 176 mph

Road tested in: Paul Frere, August 20, 1969; *Car & Driver*, January 1970
Main competitors: Lamborghini Miura S & SV; Iso Grifo 7 Liter; Maserati Ghibli SS; Monteverdi 400 SS, Hai 450 SS & GTS

« "The fundamental objective we set for ourselves was to obtain a thin, svelte car like a midengine design," Sergio Pininfarina said about the Daytona's design. "The whole idea was really a search for this sense of lightness and rake...." Needless to say, they succeeded. *Pininfarina archives*

⩔ When the production prototype Daytona (likely chassis 11795) broke cover at the 1968 Paris Auto Show, "The Pininfarina-Dino Coupe standing nearby," *Road & Track* commented, "just looked old fashioned in comparison." (The Dino seen is chassis 00114.) *LAT*

The target, though, was now Lamborghini's Miura. "Every rich and impatient man wanted one when we showed it at Geneva in 1966," Dallara recalled. And with good reason—its seminal shape looked like the proverbial million dollars, and the car cleared 170 miles per hour when tested by former Le Mans winner Paul Frere for the June 17, 1967, issue of England's *Motor*. With a number of other magazines also seeing 170-plus miles per hour, everyone expected Ferrari to respond with a mid-engine twelve-cylinder.

But that's not what happened. Enzo's Miura fighter serendipitously found its beginnings in the Pininfarina works when stylist Leonardo Fioravanti saw a rolling chassis for Maranello's new 330 GTC/GTS. "I felt it was something really rare," he recalled, "if not unique."

With his creative juices flowing, he went to work: "I wanted to faithfully follow the overall dimensions of the mechanical parts, paying extreme attention to the aerodynamics. The first drafts and more specific sketches I made later were really liked by Sergio Pininfarina. In fact, he was so positive he decided to present them to Comm. Ferrari, even if at that moment the 275 GTB/4's replacement had not yet been foreseen."

Ferrari liked what he saw, made his choice, and Fioravanti went to work on the "form plan," a type of blueprint drawing Pininfarina called "the beginning for the construction of the model first, and then the prototype."

The car would become known as the Daytona (after Ferrari's 1-2-3 sweep at the American racetrack in 1967), and two prototypes were made using 275 GTB/4 frames. The first (chassis 10287) began testing in the fall of 1967, the second (11001) in early 1968. The production car would come a year later and had a more rigid tubular chassis with a slightly longer wheelbase and wider track. The suspension was independent front and rear, with double wishbones, coil springs, telescopic shocks, and anti-roll bars. The five-speed transaxle featured limited slip, and the vacuum-assisted Girling discs were ventilated.

The interior had amenities such as air conditioning and power windows. The Daytona weighed several hundred pounds more than the 275, but this was more than offset by its sleek aerodynamic body and 352-horsepower 4,390cc V-12. The production prototype (likely chassis 11795) was first seen at the 1968 Paris Auto Show and dubbed the "anti-Miura" by England's *Motor*.

When tested by the magazines, the general automotive world had a new hypercar benchmark. Paul Frere was first with the news when his August 1969 test for *Motor* saw the Ferrari turn 0–60 in 5.8 seconds, 0–100 in 12.8, and the standing kilometer in 24.35. With a maximum speed of 176 miles per hour, he noted it was "the fastest I have ever done with a normal road car."

OPPOSITE:
The key to the Daytona's prodigious performance was a 4,390cc DOHC V-12 that produced 352 horsepower at 7,500 rpm. In traffic, the engine would easily tick along at 700 rpm even though it produced 50-plus more horsepower than the motor of its predecessor, the 275 GTB/4. Ultimate speed had indeed come a long way.

≫ The magazine tests consistently recorded the Daytona's top speed at 172–178 miles per hour. While this was the same speed as the 375 MM fifteen years earlier, the Daytona had a full leather interior, proper carpeting, air conditioning, a package tray behind the seats, and a large trunk.

« All that comfort, speed, and sophistication made the Daytona incredibly popular, and it easily outsold its Miura rival, 1,284 to 765. Indeed, the model became first 12-cylinder Ferrari to exceed 1,000 units. This is the factory in 1970, where a number of Daytonas wait to go onto the production line.
John Clinard

One year after the Daytona broke cover, the 365 GTS/4 appeared. Known as the Daytona Spyder, "When I sketched the Daytona coupe," designer Leonardo Fioravanti recalled, "I didn't plan to derive a spyder model because of stiffness and aerodynamic considerations." Instead, the spyder came from the fertile mind of Sergio Scaglietti, who found creating the car "fairly easy to do."

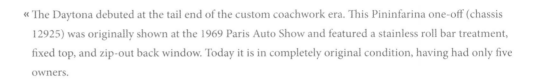

« The Daytona debuted at the tail end of the custom coachwork era. This Pininfarina one-off (chassis 12925) was originally shown at the 1969 Paris Auto Show and featured a stainless roll bar treatment, fixed top, and zip-out back window. Today it is in completely original condition, having had only five owners.

⌃ The competition wasn't sitting still as Ferrari developed the Daytona. Iso's Grifo 7-Liter had a 400-horsepower 427 under the hood, and when mated to a ZF five-speed transmission, was a legitimate top-speed contender. Iso engineer Giuseppe Caso saw 280 kilometers per hour (174 miles per hour) during testing and said the car "probably had another 10 kilometers per hour."

Subsequent tests confirmed the Miura had been knocked off its perch, and Ferrari would go on and produce 1,284 Daytona coupes and 122 Daytona Spyders between 1968 and 1973. That large production run indicates how chic ultimate speed had become. Ten years earlier such "supercars" were built in single- or very low double-digit numbers. Indeed, the only true competitor to Ferrari's 375 MM and 410 SAs in terms of speed was Mercedes' 300SLR coupe that hit 100 miles per hour in 13.7 seconds and topped out at 176–180 miles per hour, according to *Motor Racing*'s January 1957 test. But only two were made.

So what changed to create hypercar demand? The Miura was the flashpoint, coming to market at exactly the right moment. By the mid-1960s people were craving speed, and with a number of periodicals as established sources of information, a much larger number of enthusiasts, tastemakers, and consumers saw auto show coverage, race results, and road tests and wanted in on the action. Information fanned across the globe, and page counts grew tremendously as advertising budgets poured into the publications, instigating an increasing appetite for sexy machinery to put on a cover. The United States was the economic engine of the world, the joke in financial circles being, "When America sneezes, the rest of the world catches pneumonia." Europe was still booming, and Japan was in the early years of its own postwar renaissance.

The Daytona was born in this heady environment. From 1969 to 1970 other competitors upped their games: Iso had its five-speed Grifo 7 Liter, Maserati increased engine displacement in its Ghibli, upstart Monteverdi had its 400 SS and Hai, and the Miura received "S" and then "SV" upgrades. This simply added more fuel to an already hot fire, including—as will be seen—one most people didn't see simmering.

In 1971 Lamborghini threw down a serious gauntlet with its wild-looking Countach and claimed 196 miles per hour maximum. So Ferrari responded with something Lamborghini didn't possess: competition breeding.

BB = "A SPLENDID WOMAN"

By 1971, Maranello had been running mid-engine F1 cars for ten years, endurance racers for eight. Plus the 140-mile-per-hour production Dino went on sale in 1968, the magazines raved about it, and it would become Ferrari's bestseller up to that time. Pininfarina had shown several non-running, wild-looking, mid-engine hypercar studies based on racing models, each leaving the public wondering what would be next from Ferrari.

With Maranello's clientele now accustomed to mid-engines, Enzo's resistance came tumbling down. No one was more pleased than Sergio Pininfarina: "One day he told me, 'Instead of the V-12 I am trying to make a flat-12. I consider this the future of our cars.'"

Ferrari had first used the "boxer" engine in 1964 in the F1, and it was the powerplant of choice in 1971's F1 cars and the endurance racing 312 PB. "I very much liked the engine because of its space architecture," Sergio Pininfarina noted. "For years and years—and especially with Ferrari—I had to fight with a high engine and a large radiator because the engine's height [dictated] the radiator's height. . . . But with the boxer engine being lower, it made everything easier."

Ferrari's newest was the 365GT/4 BB, better known as the Berlinetta Boxer. The two men who oversaw its mechanical design and development were seasoned engineers Angelo Bellei and Giuliano de Angelis. Though a road-going flat-12 engine was unprecedented, "When we made the decision to make the Boxer, we had no qualms," Bellei told author Mel Nichols in *Ferrari Berlinetta Boxer*. "Our experience with the Dino 246 GT and the Formula 1 boxer engine combined to give us complete confidence in the path we wished to follow."

De Angelis used the 3-liter competition engine as his starting point but specified a displacement of 4.4 liters; this allowed components such as pistons and connecting rods

≈ The P6 broke cover at the 1968 Turin Show, the non-running prototype looking very close to a production car. The overall shape and proportions foretold the 365 GT4/BB (Berlinetta Boxer) that would appear at the same show three years later. *Pininfarina archives*

The Hyper Dream Cars

Former Pininfarina designer and Carrozzeria Ghia managing director Filippo Sapino was a teenager in the 1950s and recalled how his passion for cars "really started with a rally that was close by our house. The cars that went by, they looked like UFOs."

Little did the enthusiastic youth realize that one day he would design something that gave much the same impression: the extraordinary Ferrari 512 S seen at the 1969 Turin Auto Show. *Road & Track* called the startling one-off "the outstanding car of the show . . . a very bold piece of futurism."

That last word is key, for in 1965 the carrozzeria began an unforgettable run of mid-engine dream cars based on Ferrari mechanicals, and for several years onlookers wondered at what was a pure flight of fancy and what would end up on the road with them. The first to emphasize road car "potential" was the 250 LM Speciale that went to Luigi Chinetti in the United States. The following year the 365 Ps were shown at Paris and Turin and sold to Gianni Agnelli and Chinetti. (An even earlier "road car" example was the 1960 avant-garde Superfast II that went into production as the 400 SA *Aerodinamico*.)

Where the line between road car and dream car really began to blur was at the 1965 Paris Show. While the startling Dino Berlinetta Speciale was never a hypercar in today's sense of the word, that one-off set a precedent because the second, more refined Dino

Berlinetta GT seen in 1966 was the interim step to the production prototype 206 GT that debuted at Turin in 1967. The 206 then went on sale the following summer and would be Ferrari's most popular model well into the 1970s.

Dream car-to-production happened again with the P6. First shown at Turin in 1968, by 1971 the static model had morphed into the 365 GT4/BB "Berlinetta Boxer" show car. After two-plus years of development, the Boxer hit the roads in 1974 as Ferrari's hypercar model.

In addition to the 512 S, three other Pininfarina-Ferrari dream cars remained one-offs. Two were the bright yellow Dino Competizione (Frankfurt 1967), and the other was the sensational gullwing-doored P5 that broke cover at Geneva in 1968. (An Alfa "sister," the 33 Prototipo Speciale seen in 1969, was also a one-off.)

The era's wildest dream car was the last. The Modulo used 512 S underpinnings and, with a roofline hardly taller than a yardstick, looked like a flying saucer with barely visible wheels. Then societal and economic pressures put an end to such design experimentation until the 1989 Mythos (see Chapter 6). Subsequent boundary-pushers were the 2000 Rossa barchetta, and especially the 2005 Birdcage 75 (see Chapter 7). The latter was a Ferrari-based Maserati and showed how such dream cars never die, as its designer (Jason Castriota) said the inspiration was the Modulo.

In late 1969, Ferrari decided it would take on Porsche's 917 the following year. Around the time of the endurance racer's press presentation in Modena, Pininfarina displayed the avant-garde 512 S show car at the Turin Motor Show. *Pininfarina archives*

Making the 512 S look conventional was the 1970 Modulo that debuted at Geneva. Its designer was Paolo Martin, and fellow stylist Filippo Sapino recalled walking into the Pininfarina studios one weekend, where "it felt like there was a blizzard going on in there" as shavings from Martin's foam model filled the air as the designer feverishly worked away. *Pininfarina archives*

The biggest threat to Ferrari's performance and prestige crown came in 1971 from rival Lamborghini with the debut of the LP500 Countach (below). It entered production in 1974 as the LP400 Countach (above). "The car was conceived and built for one singular purpose," former Lamborghini development/ test driver Bob Wallace recalled. "To go like hell." *Richard Adatto collection*

already in production on the Daytona to be used. The chassis was a semi-monocoque design around the cabin, with tubular subframes front and rear. The front suspension had unequal length A-arms, coil springs, inclined shocks, and an anti-roll bar, while the rear also had unequal length A-arms and anti-roll bar, but dual coil springs and tube shocks.

"What we did have some doubts about the considerable weight at the rear," Bellei stated. "We knew we would have to work very hard to make sure we kept it under control. . . . With time and patience, we were able to set up a suspension that we felt was entirely satisfactory."

Sergio Pininfarina and his troops were equally busy in Turin. The BB's design starting point was the P6, a non-running concept first shown at the 1968 Turin Show. "The Berlinetta Boxer was born red with all the lower parts in black or dark gray," Pininfarina commented. "The idea was to 'cut' the car in two to make it look slender. Also, the lower part was made of a very good scratch-resistant material, while the upper material was more fragile.

"Not many people understand that when we called the Berlinetta Boxer the 'BB,' the joke was it also meant Brigette Bardot! She was a splendid woman. . . ."

Like any proper movie star, the BB arrived fashionably late for its coming-out party at the 1971 Turin Auto Show, appearing at 10:30 a.m. on press day. Deliveries began a little less than two years later, the coachwork being constructed by Modena's master panel beater Sergio Scaglietti.

Once in the hands of journalists, Ferrari found itself in an unusual quandary. Every magazine that tried the Daytona either met or exceeded its 174-miles per hour claimed top speed, but no one got close to the Boxer's quoted 188-mile-per-hour maximum. Typical was *Road & Track*'s June 1975 test: Though they called it "the fasted road car we've ever tested," it saw "only" 175. Their acceleration times also suffered (7.2 seconds

≳ » To truly understand the magnitude of the shock value Pininfarina's dream cars had, it's best to see them in a natural environment. In 1969 outside Luigi Chinetti's facility, the 1968 P5 makes everything in the background look huge, ponderous, and extremely antiquated—even a three-year-old Corvette (in the photo at right). *The Stanley Rosenthal collection at the Revs Institute for Automotive Research*

« Ferrari's response to its competitors came at the 1971 Turin Show with the mid-engine twelve-cylinder 365 GT4/BB; here the prototype is hoisted onto Pininfarina's show stand. "When speaking about a car's architecture and lines," Sergio Pininfarina reflected many years later, "the Berlinetta Boxer is one of the most significant cars Pininfarina has made." *The Edward Eves collection at the Revs Institute for Automotive Research*

365 GT4/BB, 512 BB, 512 BBi

Year(s) made: 1973–1976 (365); 1976–1981 (512 BB); 1981–1984 (BBi)

Total number produced: 387 (365); 929 (512); 1,007 (BBi)

Hypercars: all

Drivetrain: 4,390cc DOHC flat-12/4,942cc flat 12 + five-speed transmission; 360 hp @ 7,500 rpm (365); 340 hp @ 6,200 rpm (512); 340 hp @ 6,000 rpm (BBi)

Weight: 2,472 lbs (365); 3,084 lbs (512); 3,305 lbs (BBi)*

Price when new: $38,000 (365 GT4/BB)

PERFORMANCE:

0–60 mph: 6.1 seconds (365 GT4/BB); 5.1 sec (512 BB); 5.8 seconds (512 BBi)

0–100 mph: 14.1 seconds (365 GT4/BB); 12.2 seconds (512 BB); 13.4 seconds (512 BBi)

1/4 mile: 13.5 sec @ 100.5 mph (512 BB)

Top speed: estimated 176 mph (365 GT4/BB); 176.2 mph (512 BB); estimated 174 mph (512 BBi)

Road tested in: 365 GT4/BB, *Car & Driver*, January 1976; 512 BB, *Road & Track*, September 1984; 512 BBi, *CAR*, May 1982

Main competitors: Lamborghini Countach, Maserati Bora, Monteverdi Hai 450 GTS, Aston Martin V-8 Vantage

* Weights are those given by Ferrari S.p.A.

» Shown here is 365 Boxer chassis 18321. With those clean lines, it's easy to see why Sergio Pininfarina thought so highly of the car. The Boxer signified a transition for Maranello—from this point on, all its "ultimates" would have mid-engines.

to 60 was some 2 seconds off the factory figure), a slipping clutch and other maladies hampering their times.

But when a 365 Boxer was right, it flew. In *CAR*'s road test, Mel Nichols hit 60 in 5.4 seconds, 100 in 11.3, and found at 160 miles per hour there was "plenty more in hand." He stated top end was "around 290 kph" (180 miles per hour), an observation he said a BB-owning friend in Germany saw on the autobahns.

So why didn't the Boxer hit the claimed performance figures? Bellei and de Angelis stated only the first engine made the quoted 380 horsepower, at 7,000 rpm; all others were detuned 20 horsepower for improved drivability, which dropped top speed to 181 miles per hour. Unfortunately, nobody outside the inner circle knew of the change.

"TWILIGHT OF THE GODDESSES"

In total Ferrari produced 2,323 Boxers from 1973 to 1984 (387 365s, 929 512 BBs, and 1,007 512 BBs). Prior to the BB, a model's production run was typically three to five years, so that the car remained in production for more than a decade was indicative of the incredible storms Ferrari—and all the GT manufacturers—were sailing through.

The first one hit when America's EPA and DOT legislation was enacted in 1968. "Up against the wall, Ferrari lovers," was how the January 1970 article began about *Car & Driver*'s test of what they claimed was the only Daytona in the country. "The battle lines have been drawn—Washington on one side and Modena on the other—and you lose." In other words, Maranello hadn't figured out how to get the model into America legally, though it eventually would.

At the same time, a more difficult, amorphous battle was taking place on Ferrari's home soil. "In 1960, the typical middle-class family in Italy had a small Fiat 500," said Aurelio Bertocchi of Maserati. "In 1965, there were two people working in each family, and they had two cars minimum. Therefore, the employees wanted higher pay, rather than higher benefits. . . ."

Inflation started creeping into Italians' lives, and if increasing salary expectations weren't met, discontent formed. For some, that was like lighting a fuse. Part of the Italian population possessed an anarchistic, rebellious streak, and the unions, which had largely been out of power since the late 1940s, focused on this unease and began organizing strikes at the auto industry's suppliers. Sometimes the issues were real—workplace safety, for instance—other times not, but the resulting strikes affected Ferrari's (and all of Italy's automotive) production in the second half of the 1960s.

"From 1957 to 1967, these were really good working years," Sergio Scaglietti recalled. "Work was going strong, and everyone was employed. When the unions started coming into the workplace, they decided the workers shouldn't do the extra hours. That spoiled everything because the workers started to become lazy. That was the beginning, [for] it was always influence, influence, on the situation."

The simmering discontent hit another gear in 1968. Like America, Italy had its share of sit-ins on college campuses, but the occupations now became so frequent that learning went into a steep decline. Mao's *Red Book* became a bible, Communist ideals swept across Europe, and violent protests were ever more frequent. The unions noticed the trend, and college campuses became fertile recruiting grounds.

"Everything went from heaven to hell quite quickly," Scaglietti lamented. "Ferrari was really getting fed up because the workers were giving him hassles and headaches."

But Enzo found a solution in June 1969 when he sold a large portion of his firm to Fiat—all but guaranteeing its survival. "He called me and asked, 'Would you consider doing something like I'm doing?'" Scaglietti recalled. "I said, 'Give me the pen! I'm ready to sign.'

"What Mr. Ferrari did was [tell Fiat] the only way they could buy Ferrari was to buy both companies. This allowed me to retire with a good pension."

≽ Nothing feels like a Ferrari flat 12; it's as smooth as pouring water from one glass to another. And the symphony of those 12 cylinders and multithroated carburetors sucking in air while under hard acceleration is one of the best sounds in the motoring kingdom.

≽ As on its predecessor, the Daytona, Ferrari and Pininfarina gave the 365 Boxer occupants comfort and refinement. Design tricks such as hidden door handles (tucked up inside the door cutout) and pull down visors that hid in the roofline gave the interior a clean, uncluttered look.

When the 365 Boxer went on sale in 1973, the climate for exotic cars had completely changed. In October 1974, when this photo was taken, many people viewed such cars as wasteful products, or worse. This attitude caused French journalist and car aficionado Jean-Francis Held to wonder in *Automobile Year* if a "Twilight of the Goddesses" was besieging auto enthusiasts. *The Karl Ludvigsen collection at the Revs Institute for Automotive Research*

The two men's timing couldn't have been better. That year Italy's Parliament passed what became known as the "Workers' Statute," a bill that "made it very difficult for an employer to fire anybody for laziness or absenteeism," the *New York Time*'s Rome Bureau chief Paul Hoffman noted in his book, *That Fine Italian Hand*. "[It] gave unions additional powers in factories and offices [where] workers can refuse to be transferred to other jobs. If litigation ensues, pro-labor judges often decide against the employer."

Other GT manufacturers also used the same strategy of aligning themselves with what became known as "a mother ship." Maserati's savior was Citroen, De Tomaso found a temporary harbor with Ford, Iso a firm in America, and Lamborghini some Swiss investors.

But the troubles were far from over. The youth of Italy and Europe rejected capitalism so strongly that one participant recalled, "It was as if the students were saying, 'We are starting a war.'"

Soon the mindset spread beyond college campuses. "A GT car, or any display of success, became a bad thing in the eyes of the public," Piero Rivolta recalled. "This caused the marketplace to go away. Our banks started looking at us as making a wasteful product, and sales evaporated. Around this time De Tomaso told me whenever someone bought one of our cars, it was like you were giving them a cross to carry."

Beautifully summing up this dramatic perceptual change was Jean-Francis Held's "Twilight of the Goddesses" editorial in the 1972–73 issue of *Automobile Year*. "At a time when, on all sides, the automobile is being accused of killing, of polluting, and of stultifying," the mainstream French journalist noted, "those of us who carry on nurturing our passion are beginning to feel like criminals on the run. . . . Have we arrived at a crossroads? Are we experiencing, in short, the twilight of those goddesses we persist in adoring, come hell or high water?"

As if to answer that question, in October 1973 the Yom Kippur war hit, followed closely by the first oil shock. Speed limits were implemented everywhere, gas prices skyrocketed, and gas rationing became the norm. The resulting worldwide recession and loss of sales caused a number of GT constructors to go under, and industry leader Ferrari's production plunged nearly 30 percent from a then-record 1,844 in 1972 to 1,337 in 1975. As *Forbes* magazine noted at the time, "Italy, once again, is the sick man of Europe. There are bombings and strikes and a breakdown of public services. . . ."

That was especially true in Turin, home of Fiat, where the unions' power had increased so dramatically that Italy's largest corporation was now under siege. "It was total chaos," said Tom Tjaarda, an American who moved to Turin in 1959 and would eventually become head of design at Ghia in the late 1960s. "The violence forced Agnelli to hire a new kind of tough-guy manager and executive with no background in the automotive field. Their principal role was to . . . try and control the outbreaks. No one concentrated on production, only on trying to survive."

The situation became so chaotic that in 1978 former Prime Minister Aldo Moro was kidnapped by the Marxist revolutionary group Red Brigades and executed. Not long after Sergio Pininfarina got quite a start when he learned politicians weren't the extremists' only targets. At the time he was president of Turin's Industrialist Union and had returned to Turin early one morning after an overnight flight from the United States. He went directly to the office to work for a few hours, then decided to go home for lunch.

"I was driving my car slowly through the city, which was a very bad idea during this [period]," he recalled. "My car was armored, and at a certain moment I got to a red light and heard a very strong *crack crack*. I looked and saw two holes in the side glass in the door [and] thought, 'This is the period of terrorists, and they will open the door and machine gun me down!' So I ducked down, waiting for that moment."

After several heart pounding seconds, the light turned green. "I hit the accelerator!" Sergio said. "I had two guards behind me [in another car], and they followed me home, not realizing anything had happened." After a thorough police investigation, Pininfarina never knew if the shots were a warning or an assassination attempt. But the incident showed how the world had turned upside down. A man of deep integrity and keen moral principles, ten years earlier Pininfarina was embraced as a global Italian ambassador. Now, he had a bull's-eye on his back.

Then, just when it seemed life couldn't get any crazier, out of the blue came a watershed event that sent Italy's auto industry careening in a new direction.

The easiest way to distinguish the 512 BB from its 365 counterpart is two taillights per side in place of three and the wonderful "512 BB" designation on the back edge of the rear deck lid. As 512 BB production soldiered on in 1980, an event occurred in Turin that would turn Italy's automotive world upside down.

SECTION II:
THE LIGHTWEIGHT MATERIALS ERA
(1984–1997)

"The demands of mass car production are contrary to my temperament. I am mainly interested in promoting new developments [as] I should like to put something new into my cars every morning."

—Enzo Ferrari

This is perhaps the most exalted entranceway in the motoring world, leading to the inner sanctum of the factory. Back in 1981 when this photo was taken, Fiat influence was being felt, but in many ways the company still had the feel of a provincial manufacturer, ruled in large by a single, charismatic leader.

≫ When Enzo Ferrari spoke at the firm's annual lunch in 1981, he had to be pleased with his company's resilience in weathering the political and economic storms of the mid- and late 1970s. At the time, little did anyone realize how Maranello's road cars would transform over the next decade—let alone the next three.

ON OCTOBER 14, 1980, Fiat had been besieged by a crippling, month-long general strike. Sergio Pininfarina was in Turin that fateful day, chairing a tension-packed meeting with a number of distraught business leaders. As he implored them to keep up their spirits in the face of the chaos, an aide unexpectedly came into the room, approached, and quietly informed him, "Five thousand workers are marching through downtown."

A perplexed Sergio relayed the message to his cohorts, only to have the aide return several minutes later with an update: The number was now 10,000. He eventually returned two more times, saying the figure had doubled and doubled again.

"That strike was particularly brutal," recalled Tom Tjaarda, who was then director of Fiat's advanced styling department. "The company was paralyzed, so they sent everyone home. Instead of doing that, I went into the center of town. It seemed everyone else was doing the same, like the workers had reached their limits. They wanted to send a strong message to the union leaders that they had had enough of the leaders only serving themselves and not the interests of their members."

That groundswell rebellion became known as "The March of 40,000," and it broke the unions' stranglehold on labor relations. In the weeks and months that followed, the auto manufacturers and design houses were finally able to return to focusing on new products, rather than survival. "When I see our Alfa Romeo Spider from that period," Pininfarina reflected, "it reminds me of how little was done during those years. The car was introduced in the mid-1960s, and there it was, the same model, a decade and a half later."

⋩ The rear of the factory shows customer cars awaiting delivery. The year 1981 was a good one for Ferrari, as production topped 2,500 cars for the first time.

⋩ To combat increasingly stringent emissions standards, Ferrari used fuel injection across the model line in 1980–81; the injected Boxer was the 512 BBi.

While Ferrari's offerings weren't that long in the tooth, not one could be considered "new." More troubling were the emissions regulations instituted in America, Europe, and elsewhere: Compliance meant sapped power output across Ferrari's product line. For example, when the 308 GT/4 was introduced in 1973, and the GTB two years later, their 3-liter V-8s produced 255 horsepower. In model year 1979, output decreased to 230, then dropped another 25 when fuel injection was introduced in 1981.

NEW BLOOD

There was some good news, though. Since Ferrari hit its sales nadir in 1975, production had steadily increased, crossing the 2,000 mark for the first time in 1979. Additionally, Ferrari had won the F1 drivers' and constructors' crowns in both of those years.

But that didn't mean change wasn't in the air. Over the previous decades, Ferrari built its reputation on naturally aspirated twelve-cylinder powerplants in both its competition and road cars. Indeed, the twelve-cylinder engine was so ingrained in Ferrari culture that a large number of owners and enthusiasts frequently voiced, "*Real* Ferraris have twelve cylinders."

The looming change was foreshadowed in 1977 when "Renault opened up the F1 [turbo] engine era," Mauro Forghieri noted in *Forghieri on Ferrari*. "By the end of the 1979 season it had only won once, yet we clearly understood that was the right road to follow. . . . [It] was no mean feat because that kind of technology was unknown. Not only by Ferrari but all the other constructors . . . except Renault. I was probably one of the few who had some notion of it all, due to my passion for aeronautical piston and turbine engines."

Forghieri would soon get assistance in tackling the turbo issue. With retirement looming for old-time engineers Franco Rocchi and Walter Salvarani, Enzo went looking for new blood and found it with Nicola Materazzi.

Materazzi was then a forty-year-old engineer working for racecar manufacturer Osella. His experience with turbos had begun several years earlier with Lancia's mid-engine, Ferrari Dino-powered Stratos. Lancia would become a three-time world rally champion with the Stratos, and Materazzi took performance even further with the turbocharged Stratos Silhouette, a two-time Giro d'Italia winner. When Ferrari hired Materazzi in December 1979, one magazine headline noted, "The technician who knows everything about the turbo is going from Osella to Ferrari."

Enzo's timing couldn't have been better; the turbo revolution would soon extend far beyond F1. "In a world obsessed with political and financial uncertainty," Ian Fraser observed in *CAR*'s March 1980 issue, "it's comforting to know that turbocharging is as irrevocably committed to rise as the sun and the moon. [They] will indelibly be part of the language by the end of this decade [for] the facts are as simple as a bank statement: Turbocharging provides the answers that secure the future of the reciprocating internal-combustion engine."

When Materazzi joined Ferrari, Enzo's men were toiling on an all-new 1,496cc 120-degree V-6, dubbed the 126 C, for its F1 cars. "Ferrari trusted me a lot," Materazzi recalled. "Three days after I arrived he said, 'We have a problem with our turbo engine.' It seemed after a few minutes the engine broke or had some type of issue."

Over the winter they experimented with turbos from Germany's KKK and Garrett in America. A third option that Materazzi called "the miracle car" used a Comprex supercharger "that had no lag. We tried it at the Long Beach Grand Prix, but the components sat on top of the engine and weighed too much. That created real problems with the tires," and the Comprex program was dropped.

Throughout the 126 C's development, Materazzi was the main troubleshooter: "I had many of the same issues on the Stratos Silhouette, so every time something went wrong they came to me. I remember Ferrari saying, 'Cars have to be made by people who are passionate about them,'" an observation that fit the enthusiastic engineer to a tee.

« The BBi's powerplant was definitely more civilized than the carbureted 512 and especially the 365, but lacked some of the earlier models' rawness and fabulous auditory assault.

⚒ One of Ferrari's greatest concerns in the early 1980s was how emission regulations were constantly sapping power in his road cars. For instance, when the 308 GTB was introduced in 1975, it made 255 horsepower; the injected cars six years later had only 205. When Enzo learned of a 308 being thoroughly trounced on the autostrada in a race, the incident ushered in a new era of performance in Maranello.

THE BEST ENGINEER YOU'VE NEVER HEARD OF

Certain engineering names in Ferrari history carry considerable gravitas—not only for what they accomplished with Ferrari, but also with other companies. Think Gioachino Colombo, Aurelio Lampredi, Carlo Chiti, Giotto Bizzarrini, and Mauro Forghieri, and you start to get an idea.

Nicola Materazzi may not be as well known, but this effervescent, enthusiastic man is also on the list. Called "a walking encyclopedia on the automobile" by former employer Romano Artioli, Materazzi was born on January 28, 1939, in Caselle in Pittari, a small hilltop village in southwest Italy. When asked about his childhood interests, he laughed and said, "Motori! My grandfather was a doctor and part of the Automobile Club, and every time the organization sent him a magazine I read it. I started reading when I was very young, for my sister would take me to school with her."

Materazzi's father was a pediatrician who was not pleased with his son's desire to become an engineer. "He wanted me to be a doctor, like he was. But I really didn't like that idea."

Thanks to that dislike, the automotive world is that much richer. Materazzi graduated from the University of Naples with an engineering degree, then served as a teaching assistant. In 1968 he was hired by Lancia and worked in the technical department making engineering calculations. He was extremely proficient, and headed the department before moving over and becoming one of the key people in putting the legendary Stratos into production and spearheading its rallying, and later, endurance efforts with the Stratos Silhouette. When Lancia Corse and Fiat Abarth merged in 1977, he designed the single-seater Formula Fiat Abarth.

Next was a stint at Osella, where among other things he set up an M1 for BMW's Procar series. In the series' first race at Zolder in Belgium in 1979, "We had two cars—one for Bruno Giacomelli and another for Elio de Angelis. De Angelis was only twenty at the time, and started last on the grid but won against a field that included Niki Lauda, Clay Regazzoni, and Jacques Laffite. The setup on his car was completely different from the others, and at the end of the race my friends at BMW were quite amazed."

Materazzi went to Ferrari in late 1979, and worked under Mauro Forghieri, where Materazzi headed the Gestione Sportiva (Racing

Department) technical office. After co-designing and developing Ferrari's F1 cars from 1980 to 1983, he was made director of GT car engineering. In an incredibly prolific five years, Materazzi was the instrumental figure in the design and development of the Testarossa, 288 GTO, 328 GTB & GTS, the 208 Turbo, and the 412. He was also project manager for the GTO Evoluzione and F40, and conceptualized the 348 powertrain. On the side he designed and developed the Lancia LC2 Group C racer's 268C twin turbo V-8 that was based on Ferrari's 308 engine.

After so much success, Materazzi's departure from Ferrari was a shock to everyone, including the engineer. "On the F40," he said, "we worked for thirteen months straight, including Saturday and Sunday. After the presentation of the car I decided to go on holiday and went to Ferrari to say goodbye. He said, 'When you return we'll throw a party because you will become our technical director.'

"While I was away I got a call from [Fiat CEO Vittorio] Ghidella, asking if I would become director of the Alfa Corsa racing team. I said, 'No,' and when I returned to Ferrari I found another man had been made the technical director. Fiat had made the decision about my fate, and the new engineer's cars were a disaster—the 348 and others."

Materazzi left Ferrari and placed an ad in several influential magazines that stated, "Mechanical engineer with over twenty years of experience creating high-level cars and engines. Looking for a full-time job in a small, medium, or large car company in or outside Italy. But it doesn't have to be part of the Fiat Group."

After designing a MotoGP-winning motorcycle engine for Cagiva in 1990, he became technical director for Artioli's ambitious Bugatti adventure and created the EB110. When Artioli's efforts faltered, Materazzi did consulting before being made technical director for B. Engineering. There, he created the astounding 220-plus-mile-per-hour Edonis.

Though Materazzi's resume claims he retired in 2006, don't believe it. His energy level is that of a man twenty years his junior, and as he ran through a photo presentation on his laptop, he was constantly smiling, laughing, and talking a mile a minute. Clearly, the joy derived from making some of the world's great rally, GT, and hypercars hasn't left him.

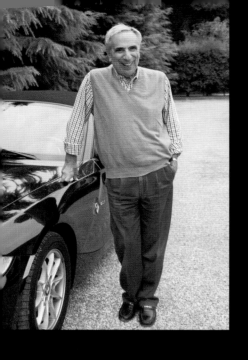

« Ferrari hired Nicola Materazzi in late 1979 to help the F1 team master the then black art of turbocharging. It wouldn't be long before the engineer's considerable skills were also applied to Ferrari's road car lineup.

» What really piqued Ferrari's interest in Materazzi was how the engineer's mastery of turbocharging could help the F1 team. Here, he's seen between Antonio Tomaini (left) and Mauro Forghieri at the 1981 Belgium Grand Prix. *Nicola Materazzi archive*

» Nicola Materazzi first learned his craft at Lancia, where he was instrumental in making the Stratos a rally winner. *LAT*

» Once Materazzi branched into road car engineering, Maranello saw a blossoming of new high-performing models. Road cars that Materazzi was instrumental in creating included the Testarossa (shown), 328 GTB and GTS, 412, and F40.

« After Materazzi left Ferrari, his portfolio would include Bugatti's EB110 and his last exotic, the astounding 225-plus-mile-per-hour Edonis. Its 3.8-liter twin-turbo V-12 approached 700 horsepower and delivered acceleration so brutal that when upshifting into third and hitting the throttle, the rear wheels broke loose.

THE GTO FLASHPOINT

The first Ferrari hypercar of the modern era would achieve its performance goals with track-bred turbo power. The 288 GTO was introduced in 1984 amid contemporary magazine assertions that it was born so Ferrari could go racing in Group B. Many articles and books have repeated this claim in the decades since the car's debut. It's surprising to learn that Materazzi, the man considered the car's father, says that is not the case. Rather, there was an event preceding Group B that acted as the true instigator for the creation of the 288.

Ever since his arrival at Ferrari he and Enzo had talked continuously, and a recurring discussion point was the 308's (and the other models') ever-declining performance. "More than once," the engineer said, "Ferrari spoke about the 'gradual gentrification or bourgeois attitude of performance' of his cars."

Materazzi recalled everything coming to a head in an episode involving "Ferrari's friend and client Pietro Barilla, who had a son named Paolo. Paolo had a BMW—I think it was a 320 or 323, perhaps drawn from Motorsport—and it easily beat a 308 on the autostrada. Enzo said, 'How is this possible, when it is a fraction of a Ferrari?' This worried

288 GTO

Year(s) made: 1984–1986
Total number produced: 272
Hypercars: all
Drivetrain: Twin-turbocharged DOHC V-8 + five-speed transmission; 400 hp @ 7,000 rpm
Weight: 2,555 lb
Price when new: $83,400

PERFORMANCE:
    0–60 mph: 5.0 seconds
    0–100 mph: 11.0 seconds
    1/4 mile: 14.1 @ 113 mph
    Top speed: 180 mph*
Road tested in: *Road & Track*, August 1984; *Road & Track*, March 1986
Main competitors: Lamborghini Countach Quattrovalvole, Ferrari Testarossa, Aston Martin V-8 Zagato

« Another sore point of Ferrari's in the early 1980s was the enormous chasm between F1 and his road cars. The F1 cars used a turbocharged V-6, but Enzo's road cars had naturally aspirated V-8s and 12s. Also troubling was lack of similarity in appearance, as found in the endurance racers just a decade earlier. *The Ove Nielsen photographic archive collection at the Revs Institute for Automotive Research*

him, and looking forward to the new models, he wanted them to return to the role that Ferrari customers rightly expected."

And so came a key conversation, what in many ways was the real flashpoint for the 288's birth, where Enzo showed the engineer a sheet of paper with projected horsepower and torque figures for a new engine derived from the 308's V-8 unit. "It was something the road car division was planning," Materazzi explained, "with peak figures of 300 horsepower and 32 kgm (231 lb-ft) of torque. Ferrari asked my opinion, and I replied the proposed values were very low, that an engine starting from a 3-liter should reach at least 400 horsepower and 48–50 kgm (347–362 lb-ft).

"Ferrari was a little surprised with my answer. But without losing heart, and almost defiantly, he said, 'If you believe you can achieve that performance, then go ahead and do it.' I objected, stating I was fully engaged in the F1 effort, and the 126 C engine.

"'Start something in the evening,' Ferrari replied. 'Then we'll find you some help.'

"That's how the 288 engine was born. It was only later, when the production cars' sales began declining, that he asked me to go to the office of the road cars and correct the situation [with] the project that would become the GTO."

THEN THE GROUP B CONNECTION

Sometime prior to that edict from Enzo, the Federation Internationale du Sport Automobile (FISA) had determined there would be three new categories for competition cars: Groups A, B, and C. Group A was modified production cars with more than 5,000 units built, Group B was any type of road car with two seats and a minimum of 200 built, and Group C was circuit-racing prototypes, of which only one had to be built. According to the Group B bible, *Group B: The Rise and Fall of Rallying's Wildest Cars,* in the summer of 1981 an announcement was made that Group B would be accepted for the World Championship Rally (WRC) in 1982 and exclusively used in 1983.

WRC piqued Materazzi's interest: "When I would speak with Ferrari, he would say, 'The difference between Formula 1 and the cars we sell is immense.' I thought to increase the performance of our GT cars we should put them in racing and liked the idea of Group B because at the time Michelotto had prepared a 308 for rallying, and someone else in France had done the same thing. Ferrari agreed, so I went ahead."

The Group B regulations stipulated naturally aspirated engines could displace up to 4 liters, while anything with forced induction had to be 40 percent smaller. While that limited a boosted engine's size to 2.85 liters, Materazzi went with that and twin turbos.

But unlike the Silhouette and the 308, the new Ferrari would have its engine mounted longitudinally, rather than transversely. "Because I wanted two turbos," Materazzi explained, "if we stayed with a transverse engine one side would have had very long exhaust pipes, while on the other they would have been very short. That would have been a disaster."

In January 1982 the engineer completed his preliminary calculations, basing the power output on the 2.85-liter V-8, using four Solex carburetors. Why Solex rather than Maranello's traditional Webers? "Because," Materazzi replied, "Ferrari really liked the Solex in his personal Renault 18 Turbo sedan."

The engineer's calculations indicated 322 horsepower at 7,000 rpm, but he was sure there was more power to come. "For eight months I worked on both the Formula 1 and GTO engines," he gleefully recalled. "I would do Formula 1 during the day and the GTO at night. For me, this was not a problem because I liked it. I really had fun working."

THE DAWN OF A NEW AGE

Around this time Ferrari contacted Pininfarina in Turin. Like Enzo's company, the carrozzeria had undergone tremendous growth since the two had begun working together three decades earlier. Pininfarina was now a design and coachbuilding powerhouse. When Sergio Pininfarina and Renzo Carli moved the company into their new Grugliasco plant in 1958, it had 36,500 square meters and 900 employees and produced nearly 5,500 car bodies per year. At the time of the GTO project, Pininfarina employed 2,400 people, had 170,000 square meters, and made around 25,000 car bodies annually.

Like Maranello, the carrozzeria was at an interesting transition point. While CAD (computer-aided design) and CAS (computer-aided styling) is found everywhere today, "When I entered into the company in the early 1970s," former Pininfarina design director Lorenzo Ramaciotti said, "there was a group of people, not a huge amount, [who] were developing software to define three-dimensional shapes. So Pininfarina was very much at the edge, for there was someone inside the company who felt that even in the specific field where we were strong, we had to be at the forefront of technology."

Sergio Pininfarina welcomed the computer: "My father had some ideas [that] were expressed with sketches. . . . From the sketches we designed the 1:1 sketch, . . . and from that a wooden model was made. The wooden model [then] had modifications that were very expensive, and very long. . . . A computer was a system through which you arrived quickly, more precisely, with no mistakes. This was a new approach to the problem, one that worked much more easily, than it used to be in the past."

The man charged with creating the GTO's shape was Pininfarina's talented and tasteful design director, Leonardo Fioravanti. He'd risen quickly through the styling department ranks so that in 1969, his fifth year at Pininfarina, he became the assistant director of the Center of Studies and Research; three years later he assumed the director's position.

Through that ascent Fioravanti designed a number of Ferraris: the 250 LM Speciale, 365 GTB/4 (Daytona), 365 Boxer, the design studies P5 (1968) and P6 (1968), the production 206/246 Dino (1967), the 365 GT4 2+2 (1972), and the 308 GTB (1975) and GTS (1977). He traveled to Maranello in early 1982, met with Ferrari and Materazzi, and was shown a modified 308 chassis that accommodated the longitudinal engine.

When Ferrari contacted Sergio Pininfarina in the early 1980s to discuss what would become the 288 GTO, the coachbuilding firm was a massive operation (below), with 2,400 employees making 25,000 car bodies a year. It was a far cry from the plant's original size, seen here under construction in 1956 (above). *Pininfarina archives*

Proud parents at the GTO's 1984 Geneva Show reveal. From the left, they are Ferrari general manager and engineer Eugenio Alzati, Paolo Pininfarina, Piero Ferrari, Ferrari chairman Giovanni Sguazzini, Sergio Pininfarina, Enzo's right-hand man Franco Gozzi, an unknown man from Ferrari's Swiss importer, engineer Nicola Materazzi, engineer Angelo Bellei, and Ferrari's Swiss importer. *Nicola Materazzi archive*

With CAD and CAS starting to play prominent roles at Pininfarina, it's surprising to learn that "We found ourselves working the same way we used to work twenty years ago," Fioravanti told automotive historian Jurgen Lewandowski in the mid-1980s for his book *Ferrari GTO.* "[It was] without the usual paraphernalia of sketches and preliminary designs. [We took] our designs straight into the workshop, like when Pininfarina built the great Berlinettas. We forgot all about computers and concentrated on the business of building the actual car."

Over the next three months the design team worked seven days a week, guided by Fioravanti's keen eye for proportions, surface development, and details. "We built a full-scale model," he told Lewandowski, "and mounted it on the first chassis. . . . When the car was completed, we did a full-scale drawing for use in the manufacturing process. Apart from that, the only drawings . . . were three or four sketches for the interior trim and a number of pilot drawings for the GTO emblem on the side of the car."

That old school methodology was in sharp contrast to the new-age materials that would be used in the GTO's construction. As in many of the advances found in Ferrari's road cars, those materials trickled down from racing.

During 1981's F1 season, England's McLaren introduced a revolutionary construction technique—a chassis made of space age, lightweight materials. In that era, an underbody air-management system known as "ground effects" dominated F1's aerodynamic thinking, so McLaren's chief designer John Barnard decided to maximize the amount of crucial "underwing" needed by "getting my chassis down to not much bigger than my driver's bum," as he so eloquently put it in *McLaren—The Cars 1964–2008.*

The breakthrough came when he visited the skunk works of the Hercules Corporation in Utah, an American firm with a division doing pioneering work in carbon composites for aerospace clients. The idea of a carbon-fiber chassis was so radical that, "It was like a big step into the unknown," McLaren's lead driver John Watson noted. "Like flying the Concorde when you've only ever flown a Boeing 707."

The concept debuted in McLaren's MP4/1 at Buenos Aires, two races after Ferrari's experimentation with the Comprex supercharger. Three months later the MP4/1 scored its first victory at the British Grand Prix at Silverstone. "It was clear that . . . a monocoque [chassis] in composite materials . . . was a new frontier," Forghieri recalled. "But we had no real engineer [for it] because that technology . . . was not available in Italy. We found one in Harvey Postlethwaite," and the affable Brit joined Maranello in the months following the British Grand Prix.

COMPOSITES, POWER, AND MORE

Ferrari first used composite technology in the 126 C2 that won the 1982 constructors' championship, and all of its subsequent F1 cars would also employ this material. The 288 GTO presented an opportunity to integrate this new material into a road car and, as Materazzi noted, "announce to the international press . . . that Ferrari was finally building the bodies of Formula 1 using this new technology."

On the GTO, the front deck lid, rear deck, and roof would be composed of Kevlar/Nomex composites, with the balance of the body made of fiberglass reinforced with Kevlar. The firewall between the engine and passenger compartments would use an aluminum honeycomb core sandwiched between two layers of a Kevlar/fiberglass composite.

While the GTO looked similar to a 308 GTB, the only body panels shared were the windshield and doors. The GTO's dimensions highlight how different they are: Its front and rear track was 4 inches wider and overall width 8 inches greater thanks to the wider tires, wheels, and beautifully integrated wheel arches.

Other engine highlights would include dual overhead cams, four valves per cylinder, new pistons cooled from beneath with an oil spray and pushing 7.6:1 compression (the

308 was 9.7:1), and a redesigned crankshaft turning in five main bearings to better handle greater loads.

In Maranello, there was much debate on which turbos to use—KKK, as employed on the F1 cars, or IHI from Japan. Both turbocharging systems were tested. The Solex carbureted/KKK-turbocharged engine (known inside Ferrari as the F114A) produced 350 horsepower at 6,500 rpm on May 22, 1982. A second engine, the F114B, was built and tested in June. It featured fuel injection and IHI turbos, and hit 370 horsepower on the dyno. IHI turbos were chosen, in part for their lack of turbo lag. Cooling was handled by an air-to-water Behr intercooler.

Materazzi was particularly proud of his solution for engine management and emissions. Working with Weber and Marelli, they devised an onboard computer that took readings of ignition advance, engine temperature, and boost to regulate fuel flow and ignition timing. The system treated each cylinder bank as if it were a standalone engine, thus each bank had its own ignition and induction system.

In final production tune, the F114B produced 400 horsepower at 7,000 rpm, which was 700 rpm below redline. The potent powerplant was mated to a trick five-speed transaxle in a magnesium alloy case that was also designed by Materazzi. The transmission's idler gears were accessible through a plate at the back to enable the quick changing of the gears to alter drive ratio. A twin-plate clutch using Formula 1-style actuation tied the transmission to the driveline.

The GTO's underpinnings represented another step forward. Though the 308's tubular frame acted as a starting point, it was heavily modified and made of high-tensile steel. A rollbar was placed behind the driver's compartment, and two tubular sideframes ran along the sides of the engine-gearbox unit. To move the center of gravity as low as possible, the GTO's powertrain sat almost 3 inches lower than the transverse unit found in the 308.

The front and rear suspension featured unequal length A-arms, coil springs over Koni shocks, and an anti-roll bar. The steering was rack and pinion with a fast 2.9 turns lock-to-lock. Wheels were lightweight modular cast alloys, 16x8 inches up front and 16x10 inches in back. The whole package was hauled to a stop by 12.2-inch ventilated disc brakes with vacuum assist.

According to Lewandowski's *Ferrari GTO*, six prototypes were built: chassis 44725 GT (a short wheelbase car that used an F114A carbureted engine with KKK turbos); 44421 GT (longer wheelbase as seen in production, IHI turbos, used in a crash test); 44727 GT (production F114B engine, brake and chassis testing); 47647 GT (production engine, car used for testing the turbos and electronics); 47649 GT (production engine, road and endurance testing); and 47711 GT (production engine, performance testing).

Development testing at Ferrari's Fiorano test track and on the road was relatively straightforward, except for "a problem that manifested itself on the initial test drive," Materazzi recalled. "This was the first time an injection system had been supplied by Weber-Marelli for use in a production car, and while testing on the road there was an abundant and violent darting of flames coming from the exhaust. While this was acceptable, and perhaps even dramatic on a competition car, it was obvious that it would not work on a road car." Modifications were made to the fuel injection system, and the issue was resolved with "no degradation in performance."

THE "INSTANT COLLECTIBLE" IS BORN

By summer 1983, word was filtering out that a new high-performance Ferrari was coming. *Road & Track*'s August issue erroneously dubbed the car the "2600 Turbo," but got it right four months later when Paul Frere wrote about "a new version of the Ferrari GTO—a homologation special . . . intended to become Ferrari's weapon in Group B."

More news broke in early 1984. "On Ferrari's superhot 308 GTO," *CAR*'s March issue noted, "some 200 . . . are being built . . . and the car is likely to be revealed at Geneva.

⌃ Once the 288 hit the road, it was clear Ferrari
was back on top of the game. The car regularly
saw 180 miles per hour and more in road tests
and had everyone singing praises.

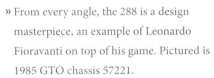

» From every angle, the 288 is a design masterpiece, an example of Leonardo Fioravanti on top of his game. Pictured is 1985 GTO chassis 57221.

≈ With those stunning looks, extremely limited production numbers, and a realistic claim on being the world's fastest production car, the 288 GTO was the first ever "instant collectible"—a motorcar that has never sold for less than its original price. The 288's resemblance to the 308 GTB is quite evident in this image. Both were Fioravanti designs, the 288's longer wheelbase, flared fenders, and upturned tail only accentuating an already beautiful shape.

≈ The 288 GTO was named after the 1962 250 GTO and paid homage to the car with the slats on the rear quarter panel.

≈ The GTO badge on the rear flank was a subtle way to let the adroit know what just blew by them.

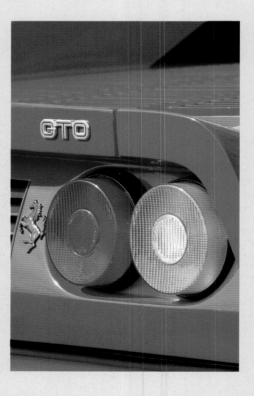

With its twin turbochargers and intercooler this powerplant delivers a stunning 350 horsepower, which should make the GTO the fastest road machine ever built behind the green gates."

That Geneva debut on February 28 was unlike any other. "There were . . . two [cars] on display," *Motorsport*'s coverage noted, "one on the manufacturer's own cramped, yellow-carpeted stand [chassis 50225], and the other revolving majestically upon a polished metal turntable on Pininfarina's typically spacious display [chassis 50253].

"Both cars were shrouded in dust covers until 5 p.m., when Ferrari unwrapped their show car for the press announcement, swamped in a struggling mass of media men and excited rubberneckers, with a sprinkling of optimistic would-be purchasers mixed in."

Indeed, so great was the scrum *Road & Track* reported that within 20 minutes after the reveal "every scrap of GTO information had been removed . . . and the beleaguered occupants were telephoning Maranello for replacements!

"[T]here was little for the would-be purchaser [who was] there. Peter Livanos, now part owner of Aston Martin Lagonda, ... paid 20,000 Swiss francs for a piece of paper saying if a car becomes available, or if they build 201, his would be the 201st!"

That launch marked the birth of the "instant collectible," those rare cars that never sell for less than the original sticker price. Brandon Lawrence was then the sales manager at Walnut Creek Ferrari in the San Francisco Bay Area, and he recalled, "At the time the GTO came out, the gray market in America (where cars not originally made for the U.S. were taken to compliance shops that converted them to U.S. specifications) was starting to get really hot. We were one of the biggest Ferrari stores in the U.S. and were told that each dealer would get one GTO. You then had to pick the client and send Ferrari a little resume on the person, saying what Ferraris he had bought in the past. Then they'd give you the okay or not and would let you know when to come and get the car.

"The store's owner was a worldly guy, and both of us sensed the GTO was going to be a hot commodity so we went around and bought ten cars from other dealers on the secondary market. The first GTO we sold went for $100,000 (the list price was $75,000), and the next was $110,000. Prices were escalating so quickly that the last two or three went for around $200,000, and they have never been that low again."

No sooner were GTOs in the hands of customers and the magazines than it was abundantly obvious that Enzo, Materazzi, Pininfarina, and company had accomplished their mission. Ferrari was *back*, and in a big way.

Road & Track's August 1984 cover story compared a 250 GTO and "the initial development prototype" (likely chassis 47649 GT, based on Lewandowski's information). Testing them was perhaps the most qualified person possible, former Ferrari F1 champion and Le Mans winner Phil Hill. He marveled at the 400-horsepower V-8, noting, "the true soul of this new GTO, just as in the original, is the engine. Ferrari purists will probably moan that this GTO doesn't have a V-12, and I admit that the sound of the 1962 GTO may be more exciting than the turbocharged V-8 . . . , but that is the only point of superiority I'm willing to concede. . . . The progressiveness of the flow of power and its remarkable control are as good or better than anything I've ever experienced."

The Ferrari accelerated to 60 miles per hour in 5 seconds, 100 in 11 and, in showing how the twin turbos built a head of steam, went through the quarter mile at 113 miles per hour. Top speed was listed at 189, the article noting that Ferrari had seen that figure at the Nardo test track in southern Italy.

"One of the most delightful aspects of the car," Hill observed, "is that despite the added horsepower and greater midrange torque, the GTO has a light, nimble feel and not the heavy, intimidating nature of, say, a Boxer or Countach. . . . To go with this lovely engine is road-holding of a very high order. The grip of the car on the road is phenomenal and noticeably increases with speed. In fact, it is difficult to induce unwanted oversteer in any gear."

⌃ The interior was comfortable and extremely functional, with no excess frills anywhere. A most pleasant place to spend a day (or more) of fast motoring.

So how did the 288 compare to the immortal 250? "In total," Hill summarized, "the new GTO is light years ahead of its 22-year-old predecessor in terms of performance, and yet it offers the option of air conditioning and Leoncavallo's *Pagliacci* in full stereo. . . . As pleased as I am to see Ferrari competing strongly in F1, I'm delighted they will once again have a Gran Turismo car with true competition potential. That's what the name GTO meant in the first place."

Several months later Hill drove a Federalized car in *Road & Track*'s "Top Speed Test" against Ferrari's new Testarossa and a four-valve Lamborghini Countach at Ohio's 7.5-mile TRC test rack. The GTO saw 180 miles per hour, gave the fiercest acceleration, and once again had the former F1 champ singing praises: "It has a super high-bred feel about it, very near to racing car agility and feel, and is not the least bit ponderous as so many of these top exotics are. And it's not just a Dino with more power; it's a beautiful package for high-speed driving."

Concurring was *CAR*'s July 1986 issue in which Roger Bell spent two full days tearing up some of Scotland's finest empty, sinuous roads. His poetic impressions are best summed up with these two observations: "One or two other cars are as finely made but, charging at full noise, none can muster the character of the GTO"; and, simply, his concluding sentence of "Here, for the moment, is the ultimate driving experience."

But how long that "moment" would last came into question a year later when a few journalists visiting Maranello came face to face with what Ferrari called the GTO Evoluzione.

⋟ The GTO's heart: the 400-horsepower F114B engine. Twin turbos, four valves per cylinder, double overhead cams, and a trick engine-management system Materazzi was especially pleased with, made for one invigorating drive.

Inside the factory in 1985, the GTO's F114B powerplants on pallets await installation. *LAT*

With 288 GTOs now having prices with two commas, the cars are rarely driven. People thus tend to forget that when they were new, they did indeed get wet as in this photo from 1985, where eight GTOs and two Testarossas await delivery. *LAT*

« All GTOs were originally painted red, except for chassis 47649, which was done in yellow. Here it is seen during Ferrari's fiftieth anniversary celebration in Modena.

⌃ In the July 1986 issue of *CAR*, Roger Bell's 288 GTO road test conclusion was succinct and to the point: "Here, for the moment, is the ultimate driving experience."

5

TARGET: 200 MPH
(1987–1992)

FERRARI'S RETURN TO HYPERCARS couldn't have gone better. The 288 GTO brought the house down at its introduction at the Geneva Show. The magazine road tests heaped praise on it. Client demand far exceeded what Ferrari's brass expected; all those factors ushered in the instant-collectible phenomenon.

THE GTO EVOLUZIONE

Though Nicola Materazzi was not surprised by the hoopla ("The people who wanted that type of car got exactly what they desired," he noted), he had little time to enjoy the euphoria. He was occupied by creating a more powerful, competition variant, the GTO Evoluzione.

"*That* was the car that was born because of Group B," he stated.

The Evoluzione's development began in mid-1983. Despite composites finding their way into Maranello's design and engineering process, Ferrari remained an engine-oriented company. Materazzi first focused his attention on the GTO's F114 B powerplant (these engines would become known as the F114 C in the Evoluzione). Group B regulations had already limited the V-8's capacity to 2.885 liters, so he increased the compression ratio slightly, from 7.6 to 7.8:1. Larger turbos with a superior airflow rate were installed and boost pressure more than doubled from 0.8 bar (11.6 psi) to 1.7 bar (24.65 psi). Valve timing and the EFI were then adjusted for reliability.

The new engine, known as the F114 CR ("R" for Rally), was first tested on October 26, 1984. It produced 530 horsepower and brought smiles to the engineering team—that output was 80 to 100 horsepower more than any car then competing in the rally championship.

While it may seem odd today to consider rallying without four-wheel drive (4WD), that did not concern Materazzi. He pointed out that Lancia won three titles with the rear-wheel drive (RWD) Stratos in the 1970s, and the WRC crown in 1983 with the RWD 037. Plus venues such as Corsica's Tour de Corse ran on tarmac. Group B also encompassed Ferrari's old stomping grounds, endurance racing. There, entrants included BMW M1s, Porsche 930s, and a handful of oddballs such as modified Porsche 928s.

Since the F114 CR had easily exceeded 500 horsepower, Materazzi knew he could turn up the wick. On went even larger IHI turbos and new inlet manifolds, the injection-ignition system was upgraded, and more aggressive cams gave the valves greater lift

» The GTO Evoluzione highlighted Materazzi's thinking: lightweight with an extremely rigid and crashworthy structure and functional interior. *John Lamm*

288 GTO EVOLUZIONE

Year(s) made: 1985–1986
Total number produced: 5-6, depending upon source
Hypercars: none; all were competition cars
Drivetrain: Twin-turbocharged 2,855cc DOHC V8 + five-speed
transmission; 650 hp @ 7,800 rpm
Weight: 2,200 lbs
Price when new: $230,000

PERFORMANCE:

0–60 mph: estimated 3.0 seconds
0–100 mph n/a
1/4 mile: n/a
Top speed: estimated 231 mph
Road tested in: *Autoweek*, January 12, 1987
Main competitors: FISA regulations

« The last key for success on the racetrack was an incredibly powerful twin turbo engine. In F114K guise, it was good for 650 horsepower at 7,800 rpm. *John Lamm*

and overlap. This variant became known as the F114 CK, the "K" for "Competizione" (Materazzi felt a "CC" designation would have been too confusing). When the engine was first tested on September 28, 1985, it produced 650 horsepower at 7,800 rpm.

By then, he and his development team were finishing up a complete reengineering and redesign of the GTO. The Evoluzione's space frame was similar to the 288's but weighed 40 percent less through the use of thinner tubes. Carbon fiber was added to key stress points, the resulting structure becoming three times more rigid than the standard GTO frame.

Materazzi said there were discussions with Pininfarina on designing the Evoluzione's body, but the responsibility ultimately fell on his shoulders. "The coefficient of drag [of the Pininfarina proposal] was too high, and I criticized this," he explained. "One of the Ferrari directors told me I was being too hard on the stylists, so I tried it myself, for the first time ever, to make sketches of possible bodies. . . . After choosing the one that seemed to be the most correct, I began producing models to develop it.

"Because this did not involve Pininfarina, we could not use their wind tunnel; therefore the prototype was tested in the Fiat Research Center wind tunnel in Orbassano, near Turin. The Cx came in at 0.293, which was quite a bit lower than what had been proposed by Pininfarina."

The Evoluziones were constructed outside of Maranello. "At that time, all of Ferrari's technicians were busy doing the 288," Materazzi said, "so as not to stop the assembly of that car, we used Giuliano Michelotto, a specialist in engines, gearboxes, and suspension. He was one of the best in Europe, and I knew him from when he was preparing the Stratos."

According to the book *Ferrari F40 LM*, on May 6, 1985, Michelotto contacted Dino Cognolato, a small coachbuilder with talented craftsmen, to "make a . . . competition evolution of the 288 GTO." The first Evoluzione was chassis 50253 GT and, like the five that followed (chassis 70167, 70205, 79887, 79888, and 79889) its coachwork was made entirely of composite materials. Materazzi's design featured a blunt nose, wide front and rear fenders, side skirts, and a plethora of vents, NACA air intakes, and louvers. On the tail was an adjustable rear wing, with a Gurney flap at its base.

Depending upon the source quoted, the Evoluzione's curb weight was 1,000 to 1,050 kilograms (2,200 to 2,310 pounds), making it at least 250 pounds lighter than the production 288. Development testing began in late 1985, Materazzi recalling the initial outing because "the test driver was late in arriving, so after a while Ferrari's managing director Giovanni Sguazzini, an engineer whom I knew from my time at Lancia, said, 'Lets go out for a lap.' The car had a wide sill on the outside of the seat, and when Sguazzini took a step over it to get out of the car, his pants ripped! I went over to my car and got him a coat, and it gave us a good laugh."

Ripped pants or not, development testing continued apace into 1986. "We had the two engines (track and rally), and the one with the most testing and evaluation was the track version," Materazzi recalled. "My friend the rally driver Lele (Raffaele) Pinto conducted the first tests at Fiorano and was very impressed. Some of the measurements from this testing told us we lacked proper tires."

Once they secured the right rubber in the form of massive Michelins, there was no doubt the Evoluzione was blisteringly fast. Period reports stated 0-to-60-mile-per-hour times of around 3.5 seconds, 0 to 125 in 10, and a top speed of 225 to 235 miles per hour.

STUNG BY A "B"

In August 1986, *CAR* ran several spy photos of an Evoluzione undergoing testing at Fiorano. "Ferrari's GTO is set to make its serious racing debut next year," the magazine reported. "According to our Italian correspondent Giancarlo Perini, it turned very quick lap times and is now likely to compete in events in both America and Europe. . . . Indications are that the factory itself may enter the GTO in certain endurance events next year to vie for Group B class honors with Porsche's 959."

And what a battle it would have been, especially on the wild and wooly WRC front. "[In] the mid-1980s rallying was more popular than F1," Group B insider Reinhard Klein reflected in the tome, *Group B: The Rise and Fall of Rallying's Wildest Cars*. But, as he and co-author (and WRC insider) John Davenport point out, "this was not liked by many of those people in power, [as] these have always been F1 people."

A quick look at YouTube verifies how widespread (and rabid) the WRC fan base was. Numerous videos show that come sun, rain, snow, dirt, ice, clean or slippery tarmac,

≋ In many ways, Group B was like watching one's favorite rock band bring down the house. Only here, the stage had numerous twists and turns, and the lead guitarist was often sliding sideways at mega-speeds or completely airborne. It was a disaster waiting to happen—and it did in 1986, when two separate fatal (and preventable) incidents ended the series. *LAT*

rows upon rows of spectators lined both sides of the courses, often in the middle of nowhere—with nothing separating them and the speeding competitors except inches (and sometimes less) of air.

"Group B was not only the ultimate in terms of performance, it also heralded the end of that adolescent thinking we all indulged in," Klein explained. "Having been there myself, I remember coming back from the Portugal Rally in 1982 and thinking, again we may have been very lucky [that] no one got killed and no one crashed into spectators."

As Klein sensed, the luck "clock" was indeed ticking. At the start of the 1985 season, the WRC had just two front-running cars (Audi's Sport Quattro S1 and Peugeot's 205 T16) with four-wheel drive. For the following year, several others (MG's Metro 6R4, Citroen's BX 4TC, Lancia's Delta S4, and Ford's RS200) joined the four-wheel-drive party, all with engines ranging from 420 to more than 500 horsepower. "For the first time," Klein and Davenport noted, "there were at least two dozen cars with the best rally drivers in the world ready to head the field and provide much more of the kind of excitement that had been building up over the previous two years. The Audi versus Lancia battles and then the Audi versus Peugeot battles were now, it seemed, going to be at least three- or even five-way battles."

Unfortunately for Ferrari, the WRC participants, and everyone else, FISA and the event organizers had continually cast a blind eye to car and track safety. The constant recklessness of the fans was stupefying (some would stand in the middle of the course, taking photos as the cars came barreling toward them), especially in the case of cars accelerating toward them propelled by 50 percent more power than had been the case at Group B's inception. The pilot's skill and the era's nascent four-wheel-drive technology struggled to keep all of that power harnessed, but lots of sideways driving was a part of the show.

The clock finally struck midnight in Portugal during the third race in 1986, when a Ford RS200 driven by local hero Joaquim Santos crested a hill to find a number of people in the road. He braked, got off course, and veered into the crowd. Three spectators were killed, and more than thirty were taken to the hospital. The drivers went on strike after this tragic incident, yet even their revolt failed to move FISA and its event organizers to address the underlying safety issues.

In early May at the Tour de Corse in Corsica, Henri Toivonen and Sergio Cresto's Lancia Delta S4 went off the road, hit a tree, and burst into flames. Authors Davenport and Klein intoned that that fatal accident gave the "F1 people" a perfect reason to kill off Group B. Effective January 1, 1987, all "special rally cars" were banned from future competition.

STUNG BY A "GRUPPE B"

That out-of-the-blue declaration "knocked out the GTO Evoluzione," Materazzi said. But Ferrari wasn't the only big-name high-performance manufacturer to get blindsided.

When Enzo and Materazzi were having their preliminary discussions in late 1981 regarding what would lead to the 288 GTO's creation, some 300 miles to the north, Porsche's head of R&D Helmuth Bott was actively engaged in a similar conversation with his engineers. Professor Bott had a running AWD 911 prototype, and the company's Group C-based mid-engine 956 under development; the latter would go on to win Le Mans, Spa, and elsewhere in 1982.

"My engineers wanted to do a mid-engine car," Bott told author Randy Leffingwell in *Porsche 911 Perfection by Design*. "I was fighting against them . . . and said 'We do so many mid-engine cars [that] we cannot learn anything.'" He thus focused his staff's attention on Group B, viewing it as an opportunity to "have it look at the future of the 911."

That they did, in an incredibly ambitious manner. A technologically advanced prototype debuted at the 1983 Frankfurt Auto Show where "Speed was the keyword," *CAR*'s coverage pointed out. "Little else seemed important as car makers vied for attention with a flood of new high-performance models."

Ferrari discovered in the fall of 1983 they weren't the only performance-oriented manufacturer eyeing Group B participation. At that year's Frankfurt Auto Show, Porsche unveiled its technology-laden Gruppe B concept car. *LAT*

It took two-plus years for the Gruppe B to morph into the 959 (foreground). Materazzi said he "very closely examined the mode of this competitor" and was looking forward to the Ferrari-Porsche confrontation in Group B. It never came, so Ferrari used the GTO Evoluzione as the basis for the F40. *John Lamm*

Porsche's prototype, dubbed the "Gruppe B," bristled with enough bells and whistles to enthrall even the geekiest enthusiast: twin turbo 400-horsepower flat-six engine featuring four valves per cylinder, DOHC, water-cooled heads, and an air-cooled block. Other highlights included a six-speed gearbox and an all-wheel-drive system with an electronically controlled torque splitting system that varied the power output between the front and rear wheels, depending upon driving conditions. Porsche announced the Gruppe B would enter production in 1984 so that it could race in 1985.

The highly publicized introduction and timetable did not go unnoticed in Maranello. "It was a concept car," Materazzi noted, "one which wanted to represent yet another evolution of the 911. I carefully examined the mode of this competitor, noting substantial differences compared to the cars I developed. While my cars were inspired by a philosophy that might be called 'rational essentialism' or, if you prefer, 'essential rationalism,' [the Porsche] was a concentration of complications, stuffed full of electronics that, at that time, were in a phase of evolution that presaged a very long development time."

As history has shown, it took several years before deliveries started, thanks in great part to engine development and supplier issues. In 1985 the Gruppe B had publicly morphed into the 959 and was formally launched at that year's Frankfurt Show. In-depth articles detailing the car's technical wizardry appeared throughout the year, noting its revised power output (now 450 horsepower) and top speed (197 miles per hour). The handful of magazines that drove production prototypes in late spring 1986 elicited declarations such as "The World's Fastest Exotic" and "Aston Martin Zagato, Ferrari GTO and Lamborghini Countach, move over. The Porsche 959 . . . is the fastest production road car of all."

HEADLINES *DO* MATTER

With Porsche dominating the speed headlines, Ferrari and Materazzi began pondering if their suddenly obsolete Group B racer could be transformed into a road car. "Ferrari had Marco Toni, one of his test drivers, try the Evoluzione," Materazzi recalled. "He went back to Ferrari and said, 'This is a great car. We have to make it.' Ferrari then said to me, 'Now you have to do a version of the car that can go on the street.' And that is how the F40 was born."

In early June 1986, just weeks after FISA's Group B announcement, Materazzi and his men were hard at work on the transformation. They focused on several key areas: which materials to use in the car's construction and making an engine and suspension suitable for both the street and track.

To determine how composites would react in real-world situations, coachbuilder Sergio Scaglietti designed and built a one-off 412 Cabriolet (chassis 65201) made of the materials. As Materazzi said, "We wanted a particularly light but stiff chassis, and the reason for choosing this particular format was we wanted to compare the car with the 2+2 coupe. We were extremely pleased with the result, for the new vehicle was several hundred pounds lighter than its predecessor, and its stiffness was improved by a factor of four or five."

"It was a fantastic car," Sergio Scaglietti beamed as he remembered the one-off. "I used it with some friends to go to the seaside, and it was very light in the rear. When you hit the accelerator, the rear wheels would break loose even though it had standard mechanicals.

"The car's total weight was 1,453 kilos (3,197 pounds, or some 800 pounds lighter than the production 412 2+2 that would appear two years later). One time near Christmas, the general manager of Ferrari spoke about this car and congratulated me on what I had done."

But there was more to Materazzi's engineering methodology than lightness and rigidity, one that ran decidedly counter to the WRC's "speed at any cost" mindset: "In my creations I have always tried to apply the *Science of Construction*, and especially the

≋ To help assess the rigidity of composite materials, Ferrari's close friend and coachbuilder Sergio Scaglietti created a one-off 412 Cabriolet. He'd been making Ferrari cars since the 1950s, and his portfolio included championship winners such as the 250 Testa Rossa and 250 GTO. To honor his role in Ferrari history, the company named the 612 Scaglietti model after him.

« The 412 Cabriolet (chassis 65201) was the perfect test bed for Ferrari to hide its intentions. The machine was the last car designed and built by Sergio Scaglietti, and the press was certain it was a 412 successor not something to assess the durability and strength of composite materials for an upcoming mid-engine hypercar.

F40

Year(s) made: 1987–1992
Total number produced: 1,311
Hypercars: all
Drivetrain: Twin-turbocharged 2,936cc DOHC V-8, 478 hp @ 7,000 rpm; five-speed transmission
Weight: 2,980 lb (U.S. specs)
Price when new: $415,000

PERFORMANCE:
    0–60 mph: 3.8 seconds
    0–100 mph: 8.0 seconds
    1/4 mile: 11.8 sec @ 124.5 mph
    Top speed: 196 mph
Road tested in: *Road & Track*, October 1991
Main competitors: Porsche 959, Bugatti EB110, Lamborghini Diablo, Cizeta V16T

⩗ The man responsible for turning the Evoluzione's design into an acceptable road car was longtime Pininfarina stylist Aldo Brovarone. He began working for the company in the early 1950s and was responsible for the 365 Ps seen in Chapter 3. The F40 was his last design before retiring.

Here an F40 prototype is put through its paces at Ferrari's test track in nearby Fiorano. Behind the wheel is test driver Dario Benuzzi. *Pininfarina archives*

Once Brovarone and Pininfarina design chief Leonardo Fioravanti had converted the 288 GTO Evoluzione into the F40, the shape underwent testing in Pininfarina's wind tunnel to refine it further. *Pininfarina archives*

⌃ Dario Benuzzi put his imprint on many of Ferrari's most memorable road cars. He started working in Maranello in 1971 and became so adept at driving and analyzing a car's dynamics that he told *Top Gear* he could "spot if there's a problem—and what it is—in 500 meters." *Matt Stone*

Ethics of Construction. By this, I mean I would not barter some insignificant increase in the stiffness of a frame with a possible decrease of protection for the driver or passenger of the car."

He thus made an incredibly robust passenger cell composed of a steel space frame structure of thick tubes that ran the length of each side of the car. Crossmember tubes linked the outer tubes at the front and rear bulkheads, while another, large diameter tube ran across the center to join them. Steel bracing resided within the A-pillars, behind the passenger compartment, and across the roof. Composite panels made with carbon fiber, Kevlar or Nomex were glued to the frame with advanced adhesives, the resulting structure increasing chassis strength and stiffness by 300 percent while weighing 20 percent less than a traditional space frame.

The steering was unassisted rack and pinion, and the suspension benefited from Ferrari's on-track experience. Unequal length upper and lower A-arms were found front and rear, as were an anti-roll bar and coil springs over Koni shocks. The spring and shock units rose inward from the top of an alloy upright; in front they sat between the wishbones. The shocks had an adjustable ride height that was achieved by a hydraulic system built into the shocks and was powered by an electronically driven pump.

The brakes were 12.9-inch Brembo discs with an aluminum core sandwiched between cast-iron friction surfaces. The discs were ventilated and drilled for extra cooling. There was no servo assist, and separate hydraulics were used for the front and rear brakes.

The lightweight Speedline alloy wheels had the traditional five-spoke pattern and were 8x17 inches in front and 13x17 inches in back. The wheels used a single locking nut and securing pin, as was typical in competition cars.

Materazzi noted that the F114 CR engine (the 530-horsepower rally version) was their starting point. The new powerplant would become known as the F120A, and it had a slightly larger bore (82mm vs. 80mm) and shorter stroke (69.5mm vs. 71mm) than the 288 GTO/Evoluzione's V-8. This brought total displacement to 2,936cc. The engine was all-alloy construction with DOHC, four valves per cylinder, and separate Weber-Marelli EFI and ignition systems for each bank of cylinders. The IHI turbos ran higher boost than the 288 (15.95 psi [1.1 bar] vs. 11.6 psi [0.8 bar]) and air-to-air intercoolers. Final power output jumped from the GTO's 400 to 478-horsepower at 7,000 rpm for the F40. Torque was up as well at 425 lb-ft at 4,000 rpm vs. 366 at 3,800 for the GTO.

The engine redlined at 7,750 rpm, and all the power went to the rear wheels via a five-speed gearbox that was inline with the engine and positioned behind it. A limited slip differential put the power to the ground.

The F40's design was done at Pininfarina. "Ferrari came to us because they felt the car was impossible to sell without a decent style," explained Lorenzo Ramaciotti, who was the design studio's deputy manager. "The performance of the Evoluzione was a reference in terms of downforce, drag, and cooling flow. We did not have much of a budget to spend to refine performance, so we stayed with the specifications of the Evoluzione."

About the inspiration for the design, "we had references from the long distance racers of the time," Ramaciotti said. "In fact, I had personal experience with the Lancia Montecarlo Turbo and used that as a starting point for the design of the car. The use of wings and scoops was standard for the supercars of that period, for in the mid-'80s we all understood the importance of downforce for supercars."

Also influencing the look was a strong social trend, one that Ramaciotti felt benefited Ferrari: "The match between the extrovert and flamboyant look like the Testarossa, and the demand for a more hedonistic social behavior all over the world and not only in Italy was a very lucky coincidence. It boosted the sales of Ferrari and started its booming fortunes. This was particularly felt in Italy, where the 1970s was a decade that was not only economically stressed, but even more stressed socially.

> " In my creations I have always tried to apply the *Science of Construction*, and especially the *Ethics of Construction*. . . . I would not barter some insignificant increase in [frame] stiffness with a possible decrease in protection for the driver or passenger."
>
> — Nicola Materazzi

"I say that when we design cars we try to express the best shape we can imagine. We are living in our times, so even in a nonconscientious way, we put in them the flavor of the times."

Charged with "expressing" the F40's look on paper was longtime Pininfarina designer Aldo Brovarone. The job took about a week, and he said he "studied the GTO Evoluzione before starting to develop ideas. After some consulting and suggestions from Ing. Fioravanti I had a complete free hand to style this Ferrari. In practical terms my work was to become a plastic surgeon: The central part of the roof and the side section remained the same, and I redesigned the front and rear part.

"The form of the new rear view came from my desire to integrate the wing with the side design of the car, for the wing support design of the GTO was, in truth, rather ugly. My design united this with the rear fender, using a steep angle to avoid an excessive surface area that would have increased wind resistance."

The job was Brovanone's last before retiring. "Ing. Fioravanti was a good design director," he happily recalled. "He carried out all the follow-up design and development work on the F40. With our designers he was really a collaborator rather than a boss. . . . We always remained very good friends."

After Pininfarina created a full-scale foam model, it traveled to subcontractor Cognolato in Padova. There the craftsmen hand made an aluminum-body master model; from this molds were made for the F40's composite body.

"For the construction of the prototypes," Materazzi said, "the F40 had a process that was completely different because we used outside companies. All the carbon parts that were not structural but torsional in function were supplied externally. A lot of effort went into learning what was necessary to design an entire shell made in composite materials."

The first running F40 prototype (likely chassis 70167, the second Evolutione that was converted into an F40) was in Maranello by February 1987. Initial testing at Fiorano went quite smoothly, Materazzi stating the first cars "didn't encounter any real unknowns, much of this due to the experience gained from the Evoluzione."

Updates were minor. In the book "Ferrari F40," Maurizio Manfredini, head of the experimental department, noted the team had to "lower brake temperatures and remove vibration," make "minor aerodynamic modifications . . . to the front of the spoiler and the rear of the undertray," and "find the best tires to put the power on the road."

After preliminary testing wrapped up, the car was driven 550 miles to the famed Nardo test track in the southwestern corner of the "heel" of Italy. The oval track was nearly 8 miles long, and a favorite for high-speed testing because the continual banked curve's width (50-plus feet and four lanes) meant at 150 miles per hour in the fourth lane, a driver didn't have to turn the steering wheel.

Piloting the prototype to Nardo and on the track was the legendary Dario Benuzzi, a man *Top Gear* so aptly called "a decisive, precise and staggeringly fast driver."

"The car was very pleasant, and easy to drive over such a long trip," he told author Mark Hughes in *Ferrari F40*. "Around Nardo, it is also comfortable to drive at its maximum. It is very stable, and the downforce is so good that it feels strongly attached to the road. Although its maximum speed is a lot higher than a Testarossa, I find it much easier to handle."

In fact, testing went so smoothly that the most disruptive occurrence happened in April when Fiat honcho Vittorio Ghidella paid a visit to Maranello. "The initial F40 programs were geared for its unveiling at Frankfurt in September," Materazzi remembered. "But the Fiat Group was to present the new Alfa Romeo 164 (sedan), and they did not want the press and other media to be distracted by this revolutionary Ferrari. I assessed the project's progress, which we discussed in the presence of Ferrari, and suggested the presentation be moved to immediately after Frankfurt. But Ferrari was adamant about July, forcing me and all my men to perform a real 'tour de force,' where everyone was working Saturday and Sunday."

HIDING IN PLAIN SIGHT

The next three months were incredibly hectic, for the team needed two running prototypes within weeks. Amazingly, the press hadn't realized that Ferrari was on the verge of a new hypercar, even though Maranello's brass had dropped subtle hints. For instance, in *Autocar*'s January 28, 1987, driving impressions of the Evoluzione, Ferrari managing director Giovanni Razelli said, "Light bodywork materials such as those used in the Evoluzione and the GTO will find their way in our production cars."

≫ How times have changed: In July 1987, with Ferrari locked in a speed battle with Porsche's 959, the presentation was decidedly low key and honestly straightforward when compared to today's massive productions. The car sitting under the cover was the calm before the storm—in more ways than one. *LAT*

» After a brief summary speech by Enzo Ferrari, the cover was whisked off to reveal F40 prototype chassis 73015. "Everything went nuts when that happened," remembered John Lamm, one of the journalists present. *Leonardo Fioravanti archive*

⌃ Key personnel who spoke at the F40 reveal were (from the left) Nicola Materazzi, Leonardo Fioravanti, Enzo Ferrari, and Ferrari president Giovanni Razelli; standing behind Ferrari and Razelli is the ever-present Franco Gozzi. This was the last model introduction done by Enzo, who passed away in August 1988. *Nicola Materazzi archive*

» A marvelous photo by John Lamm captures the essence of the F40 reveal: spotlight on the car as people swarm about, Ferrari and his men sitting quietly at the table. An easy way to tell the prototypes from the production cars is the five slats below the wing (the production cars had four). *John Lamm*

Prominent American photojournalist John Lamm was one who didn't connect the dots. "One day in 1987, we were working at Ferrari's Fiorano test track on a story for *Road & Track* when a rather amazing looking 288 GTO showed up in the pits," he observed in his book, *Supercar Revolution*. "Was it a competition car? Sure looked the part, but Ferrari told me its correct name was GTO Evoluzione. . . . What I didn't realize until later was I was looking at one of the prototypes for the soon-to-be introduced F40."

Not too long after that, *Road & Track* and a select number of other publications received a very low-key invitation from Maranello. "If you happen to be in Modena this particular day," it stated, "you might be interested in seeing a new car."

American journalist Ken Gross was at the launch, covering it for *Automobile* magazine. "It was a big deal," he recalled, "with most of the key magazines worldwide represented. We knew it was the successor to the 288 GTO, but we did not know the new name. It was the 40th anniversary of Ferrari, and we were told we'd be seeing a commemorative model that celebrated all they had learned in 40 years of car production and racing.

"There was palpable excitement beforehand; we felt we were seeing history unfold. People thought (and expressed) that this might be the last time we'd see Enzo Ferrari. For most of us, it was."

After the reveal, F40 73015 was taken to nearby Fiorano for some hot laps, where another photographic scrum ensued. "The PR guy, Pietro di Franchi, told us we would go off with the car afterwards," remembered Lamm, who was working for *Road & Track* at the time. "We did, taking up into the hills behind Maranello. It was magic." *John Lamm*

"I was the guy from *Road & Track* that was assigned to go," Lamm said. "We knew there was something special to be shown, and the night before the presentation the bar of Modena's Fini Hotel sounded like a session of the United Nations, with many languages being heard. No one was about to miss this meeting.

"Ferrari added to the mystery the next day when we entered the hall to find a lovely draped shape sitting under the spotlights. It was quite hot as we waited [and] finally, a side door opened, adding a corner of bright light to the dimly lit hall. Mr. Ferrari . . . was helped to the table, and that's when the place went nuts."

Once the ruckus died down, a frail-looking Enzo spoke through a translator: "On July 6 last year I asked my research department to look into the feasibility of building an exceptionally powerful sports car incorporating the very latest developments in engine and assembly technology. Only six days later, the directors gave the project their blessing, and now, barely a year later, the finished car stands before you. It's 40 years since the first Ferrari left the factory. On March 12, 1947, the 125 S was presented to the public for the first time. On May 25 of the same year, Franco Cortese won the Rome Grand Prix, driving a Ferrari. Now, forty years later, the Ferrari F40 demonstrates that Ferrari is still a byword for technological excellence and exceptional performance."

The F40's cover was then whisked off, and "there was a collective gasp, then everyone cheered," Gross said. Once the thunderous applause and rapid-fire flashbulbs died down, Fioravanti, Razelli, and Materazzi spent time going over the highlights of Maranello's newest.

"Then they took the car to Fiorano for photography, and it was a clusterf*&# [due to all the photographers on hand]," Lamm recalled. "The PR guy, Pietro de Franchi, told us to wait, that we would go off with the car afterwards. We did, taking it up into the hills behind Maranello for a late-day shoot. It was magic."

WITHOUT THE FATHER

By now, a bleary-eyed Nicola Materazzi was ready for a break. "We worked for thirteen months straight, weekends included," he recalled. "After the car's presentation I decided to go on holiday and went to Ferrari to say goodbye. He said, 'When you return we'll throw a party because you will become our technical director.'"

Then came a twist of fate suited for an Italian opera. While on vacation Materazzi received a phone call from Fiat's CEO Vittorio Ghidella, asking him to become director of the Alfa Corsa racing team. Materazzi loved his work at Ferrari, and replied thanks, but no thanks.

"When I returned to Ferrari after my vacation," he sadly recounted, "I found another man was the technical director. Fiat had made the decision about my fate, not Ferrari. The new engineer's cars ended up being a disaster—the 348 and others."

The one regret Materazzi had was "the fact that we could not meet the 959 on the track. Based on my five years of experience with the Reparto Corse Lancia, I was certain that the GTO, with two-wheel drive, would have had a good chance of success against the 959—just give a look at the gear ratios and the speed through the gears."

He wasn't alone. Since the F40's unveiling, enthusiasts, potential owners, and the motoring press wondered who made the fastest car. Though Porsche began delivering 959s in April 1987, getting the two side-by-side, hooking them up with instruments, and taking accurate recordings wasn't easy. In *Road & Track*'s October 1991 test of a Federalized F40, the car hit 60 in 3.8 seconds, 100 in 8.0, and blasted through the quarter mile in 11.8 seconds at 124.5 miles per hour, which was 5 miles per hour shy of redline in third gear. Top recorded speed was 196 miles per hour at 450 rpm below redline, so theoretically there was more to go.

For matter of comparison, four years earlier in Europe, Paul Frere tried both a 959 Deluxe and the lighter Sport model in *R&T*'s July 1987 issue. Sixty came up in 4.0/3.6

≫ The engine compartment of F40 prototype chassis 73015, replete with large Behr intercoolers. The production cars' engine bay would look very similar. *John Lamm*

F40 production commenced in 1988, approximately a year after the July 1987 introduction. This photo was taken in 1989 at Carrozzeria Scaglietti, where a worker is adding composite structural supports to the F40's steel-tube frame. Scaglietti became a subsidiary of Ferrari in 1969 as part of Enzo's deal with Fiat.
Michael Dregni

seconds, 100 in 9.4/8.2, and the quarter mile in 12.4 at 116 miles per hours, 11.9 at 119.5 miles per hour. With the tests four years and nine time zones apart, how much can be read into which was the better performer is suspect at best.

The most telling answer—and one that Materazzi did not know of until this book's author informed him—was in *Autocar*'s June 8, 1988, issue. The magazine teamed up with 959 owner Walter Brun, a successful Group C racer who had podium finishes at Le Mans in Porsches. They drove his personal 959 to Maranello and, as author Malte Jurgens observed, "There is a sparkle in [Brun's] eye for what lay in store: We were going to be the first to find out what happens when you pit the technical might of the Porsche 959 against the comparatively simple but even more potent Ferrari F40."

Joining Brun in testing the two cars was Ferrari's Formula 1 hot shoe, Gerhard Berger, and his assessment of the F40 was straightforward: "It is really an act of provocation to offer a car with so much power and so little weight to an ordinary driver." He also noted, "liking hard suspensions" and "the F40 has the best road chassis I've ever driven." Brun had a similar opinion, observing, "The F40 runs nearly as well as my Porsche 962. With racing tires and some fine tuning it would easily run at the front of category C2."

On Fiorano, author Jurgens marveled, "The Ferrari seemed to suck in the tarmac ribbon of the test track so fast that at the end of the short straight, the Jeager speedometer was showing about 160 miles per hour. The 959's speedometer would show 138 at the same point."

In the end, Berger felt the 959 was "at least ten seconds behind the F40 . . . at Fiorano," while "Brun would admit to six." The two drivers concluded that "the 959 is a civilized two-plus-two with amazing driving performance," and "the F40 is a racing car for the road."

⊻ The interior of F40 prototype 73015. The only real difference between it and the production cars was the door panel would change, as the production cars had roll-up windows. *John Lamm*

» At Carrozzeria Scaglietti in 1989, a worker
is smoothing the composite body that came
from Cognolato in Padova. The body would
then be primed and painted in Maranello.
Michael Dregni

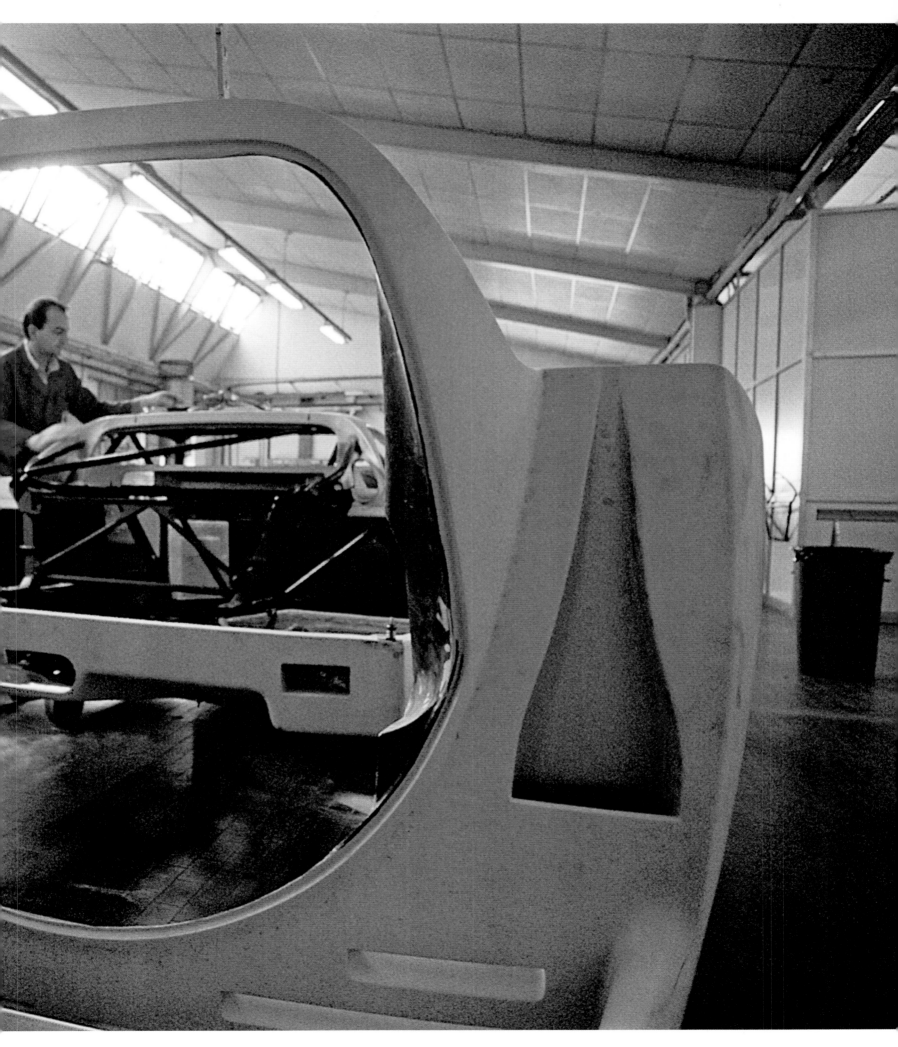

≈ The start of the dedicated F40 production line in 1989. The bodies have come from Ferrari's paint shop, without the top section of the rear wing. It would be put on at the end of the production process. *Michael Dregni*

≋ Along the production line in 1989, an F40 undergoes scrutiny before moving to the next station. *Michael Dregni*

CAR's July 1988 959 vs. F40 cover story had much the same conclusion. About the Ferrari: "The F40s performance is absolutely astounding," authors Georg Kacher and Gavin Green wrote. "It is a road car unlike any other: a machine that will venture onto the street and dismiss all other performance machines built before it with arrogant, disdainful ease. . . ."

"Forget about a quiet little Sunday drive down to the pub [for] this car is incapable of generating relaxation. It breeds anxiety and tension, yet it also delivers more sheer exhilaration than any car ever built, owing to its speed, and its noise."

Much like *Autocar*, they felt "the 959 is the better car" while "the F40 is supreme for breath-taking excitement."

BACK ON TRACK

Autocar and a number of other tests alluded to the F40 being a racecar for the street. Materazzi said that is exactly how the model was engineered from Day One: to go racing. The F40 thus marked Ferrari's return to producing true dual-purpose machines—cars equally at home on the road and track.

The gifted engineer was quite cognizant of Maranello's road and race history, as it was a frequent thread in his talks with Enzo Ferrari. "He was convinced that F1 races had become quite sterile and was concerned about the absence of experience gained on the track in the GT series," the engineer said. "F1 cars had become a world apart and had dug an extensive moat that was far away from the GT series.

"Unfortunately, the arrival of a member from Turin—read Fiat—had forced him in 1974 to stop producing the types of sports cars intended for endurance racing. Not only did [endurance racing] offer proof of [a car's] validity from a technical point of view, but it allowed us to collect necessary and valuable data that would lead to the car's improvement. In the end . . . Ferrari understood the importance of this and reaffirmed it with his brief words at the introduction of the F40.

continued on page 140

≈ At the end of the production line in 1989, a worker glues the Plexiglas rear window in place. This photo shows the magnitude of demand for F40s; the 1,311 produced vastly exceeded Enzo Ferrari's desire of keeping it at 400. *Michael Dregni*

» This is F40 chassis 84887, which is in completely original condition with very few miles on the odometer. "It was very easy for me to design this vehicle," recalled Aldo Brovarone. "It was sufficient to apply the classis forms and proportions in the traditional Pininfarina tradition."

The interior illustrates engineer Nicola Materazzi's desire for simplicity and light weight. The sills, seats, and door panels all show exposed carbon fiber. He wanted to avoid the electronic "complications" found on Porsche's 959.

≫ ↗ Perhaps the F40's most memorable design detail was the rear wing. It has four slats, compared to the prototype's five.

≫ F40 engine compartments don't come much more original than this. Visually, it appears quite similar to the Evoluzione's F114 V-8, as that was the starting point. For the F40, the stroke was shortened to 69.5mm, while a larger bore bumped displacement to 2,936cc. Final output was 478 horsepower at 7,000 rpm.

continued from page 135

"There, I also reaffirmed it by listing features introduced in the series production car, those that had been designed and planned for a track version."

The engineer then ticked them off: a fuel cell as required in competition; non-synchro gearbox (which was optional on the road cars); a robust, crash-resistant cockpit; cams profiled to "optimize power"; IHI turbos that could stand the rigors of racing; a suspension system that automatically lowered at 120 kilometers per hour (75 miles per hour); and an adjustable front spoiler and rear wing.

TARGET: LE MANS

When the F40 was introduced, the FIA had been "in favor of admitting GT cars to the Sports-Prototype World Championship," according to the book *Ferrari F40 LM*. Enzo Ferrari and FIA president Jean-Marie Balestre were talking about the topic, so on September 30, 1987, Giuliano Michelotto contacted Dino Cognolato to inform him "Ferrari envisaged racing the F40."

After Enzo passed away in August 1988, the endurance racing torch was carried by Ferrari president Giovanni Razelli and Daniel Marin, a 25-year veteran with Ferrari's French importer, Charles Pozzi. Pozzi's Ferrari France organization had previously campaigned a competition Daytona and a 512 BB LM and rallied a 308 GTB built by Michelotto.

On April 14, 1988, a frame was delivered to Michelotto in Padova. The first structural rigidity tests were carried out on May 3 and showed a 65 percent increase over the Evoluzione's frame. During the month of June coachwork was placed on the frame, and a second chassis underwent rigidity testing on June 27.

By now Michelotto and his team were working closely with 34-year-old Ferrari engineer Luigi Dindo on the mechanicals. Dindo used Ferrari's CAD (computer-aided design) to expedite the research and prototyping process, and upgrades included reinforcing the transmission's gears and adapting the suspension and mounting points for the rigors of competition.

⌃ F40 chassis 84887 is a marvelous example for highlighting originality. Red paint that was applied at the factory still sits atop the shock tower bolts and numerous other pieces under the front and rear deck lids.

F40 LM, F40 GT, F40 GTE

Year(s) made: 1989–1994 (LM); 1992–1994 (GT); 1994–1996 (GTE)
Total number produced: 20 (LM); 7 (GT); 7 (GTE)
Hypercars: none; all competition cars
Drivetrain: Twin-turbocharged DOHC V-8, 720 hp @ 7,500 rpm,
five-speed manual transmission (LM), 590 hp @ 7,800 rpm, five-speed manual transmission (F40 GT); 720 hp + six-speed sequential transmission (GTE)
Weight: 2,310 lb (F40 LM & GT); 2,250 lb (F40 GTE)
Price when new: n/a

PERFORMANCE:
0–60 mph: n/a
0–100 mph: n/a
1/4 mile: n/a
Top speed: 228 mph (LM, measured at Nardo)
Road tested in: n/a
Main competitors: Jaguar XJ220, McLaren F1 GTR, Venturi 600 LM, Porsche 3.8 RSR

« Ferrari formally returned to endurance racing in October 1989, when F40 LM chassis 79890 competed in America's IMSA series at Laguna Seca (shown). The car had seen 228 miles per hour in testing and jumped into the lead, eventually finishing third because of tire issues. *LAT*

The sister car to F40 LM chassis 79890 was 79891. A major visual difference between the two was 79891's fixed headlights.

Modifications to the powertrain included different cylinder heads, lighter connecting rods and pistons, different valves and springs, upgraded turbos with different waste gates, and flywheel and clutch. The engine's redline was now 8,000 rpm, with peak power output of 720 horsepower coming 500 rpm below that.

The first F40 LM (chassis 79890) underwent initial testing at Fiorano on October 25, 1988. Dario Benuzzi turned a 1:23.1 lap time, and three sessions later, on November 24, lap time had dropped by nearly four seconds. The team now alternated testing between Fiorano, Mugello, and Nardo, and on February 12, 1989, Benuzzi saw 367 kilometers per hour (228 miles per hour) at the high-speed track. He then recorded a 1:18 lap at Fiorano three days later.

The development team's target was the proposed GTC category in FISA's World Sports Prototype Championship. But there wasn't enough interest from other manufacturers, so the class was scratched. "We then thought about entering our cars in certain races, particularly Le Mans," Ferrari France's Jean Sage recalled in *Ferrari F40 LM*. "But the rule imposed by Bernie Eccelstone required everyone to take part in all the races," so the idea was dropped.

IMSA TO THE RESCUE

In the spring of 1989 the team acquired a new target, but it wasn't until late July that they got the green light to race in America's IMSA series. "We had to learn everything from scratch, with a newly formed Franco-Italian team operating thousands of kilometers from its home base," Marin remembered in *Ferrari F40 LM*. "We installed an aerial at Schneider's, the Ferrari concessionaire in Detroit [and] had to resolve a lot of logistical problems. For example, the petrol supplied by AGIP arrived from San Francisco. We also had to be careful about our stock of tires, which were first supplied by Pirelli.

"In order to keep up morale, we had food sent out by container so that the mechanics could have their pasta with Parmesan cheese, as well as the type of mineral water they were used to drinking."

The U.S. shakedown took place at Heartland Park in Topeka, Kansas. Then the team traveled to California's Monterey peninsula to participate in the season's second to last race at Laguna Seca.

Piloting chassis 79890 was F3000 champion and future Ferrari F1 star, Jean Alesi. "Given that Alesi didn't know the circuit," Marin recalled, "the speed with which he got into the action was truly astonishing. During the qualifying session, he had his qualifying time almost at once, after just three or four laps. . . . Only [Hans] Stuck managed to beat him in the end."

Once the green flag dropped, the LM jumped into the lead, holding it until its Pirelli tires started to degrade. Alesi ended up finishing third, behind Stuck and IMSA stalwart Hurley Haywood. The following week at Del Mar, the LM was a DNF due to ignition problems.

The crew returned to Europe and continued testing. By now Piero Fusaro had replaced Giovanni Razelli as Ferrari's managing director and put his support behind the program. The Japanese company Artsport 0123 became a sponsor, and a second LM (chassis 79891) began testing in late March at Vallelunga, driven by the experienced F1 and endurance racing driver and engineer Jean-Pierre Jabouille.

With Goodyear now supplying the tires, the team returned to America for IMSA's seventh race of the season at Heartland Park. There, chassis 79891 was retired with turbocharger problems. Both 79890 and 79891 then competed for the balance of the year, the latter running in five of the six races they entered, 79890 in three. Highlights included 79891's third OA at Mid-Ohio and second OA at Mosport in Canada, and a second OA at Road America and Watkins Glen for 79890. Both cars lead races and were driven by names such as the aforementioned Jabouille, F1 driver Jacques Laffite, and endurance racer and France F2 and F3 champion Michel Ferte.

⌃ How to make a spartan interior even more bare: prep it for racing! The seats are deep fixed buckets, and the dash has a single panel with a bar tachometer rather than individual instrument binnacles.

« As on the production F40s, the F40 LMs used a transparent Perspex rear decklid. Most noticeable are the extra-large intercoolers, which help bump power output to 720.

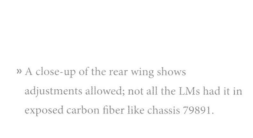

» A close-up of the rear wing shows adjustments allowed; not all the LMs had it in exposed carbon fiber like chassis 79891.

⩲ F40 LM 79891 placed 3rd overall at Mid-Ohio and 2nd overall at Mosport in 1990. The F40's remarkable silhouette is punctuated at the back by a large, adjustable carbon fiber wing.

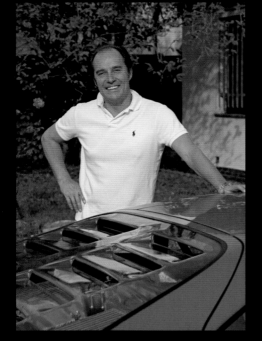

≋ Franco Meiners is one of the few people remaining who can say Enzo Ferrari personally hired him. Meiners worked in Maranello for 10 years, helping organize the factory Ferrari Club, being a key person in the birth of the F40 GT and GTE, then almost running a team of F50 GTs for Benetton.

≋ The F40 GTE was the most potent of all the F40s, having a 3.6-liter, 720-horsepower V-8. Developed and built by Ferrari subcontractor Michelotto, the model debuted in 1994 at Jarama with chassis 90001. Shown here leading the race, 90001 would retire with mechanical problems. *LAT*

Franco Meiners

Franco Meiners has lived a life most enthusiasts dream about. Born in 1965, "my father was responsible for Ford of Europe," he said. "He had a Mangusta and other things like that, so you could say I was born into cars."

Meiners' talent behind the wheel became apparent at a young age while the family vacationed in Cyprus. They went to a go-kart track with some friends, and seven-year-old Franco promptly lapped everyone. They switched cars, and Franco did it again.

"That's how it started," he said. "When I got older, I didn't have enough money to race a truly competitive car, so friends with historic cars asked me to drive. I was winning a lot, which made my friends happy, as it was possible to sell a car for more than it was worth."

As he became known throughout Europe's historic racing community, he befriended Sergio Cassano, president of Ferrari Club Italia; that relationship led to his hiring by Enzo Ferrari in 1986. "Cassano and Corrado Cupellini put me in touch with him," Meiners recalled. Then just twenty-one and a headstrong, hard charger, "I negotiated directly with Enzo Ferrari. He wanted me to be an employee, while I wanted to be a consultant. We danced

around for a few months, and he finally hired me as a consultant.

"What really impressed me about him was his ability to evaluate people very well. Right away he would understand the quality of the person, and who was able to give what."

On why Ferrari had an interest in seeing his old racecars back in action, "those were the last years of his life, and I think he was thinking about what would stay 'alive,' how would people remember them. He had this idea of not only putting together a modern racing team [with the Ferrari Club], but also a historic racing team."

Meiners played an instrumental role in the club's formation, the F40 GT and GTE programs, as well as the 348 GT and LM. He remained at Ferrari through 1996, when he left the company and moved to Paris. One of his favorite memories occurred in 1997 at Ferrari's fiftieth anniversary celebration where he turned the fastest lap at Fiorano in a 250 SWB, beating all the owners of modern machinery.

Today he is as active as ever in the automotive scene, historic racing, and dealing in high-end collector cars from his discreet facility in northern Italy.

While the effort held much promise it didn't have the full might of the Ferrari organization behind it and ended after the second season. "The car was simultaneously fabulous and a little crazy," Ferte recounted in *Ferrari F40 LM*. "With the turbos, the power arrived a little brutally, and I ended up at right angles to the track more than once. . . . I glimpsed the potential, which was not entirely exploited; there was still a lot of things which could have been done to improve the car. I would like to have continued racing with that team. We got on very well together."

Ferrari and Michelotto would go on to make another 18 LMs for favored clients, but not one of those cars raced.

A NEW TORCH BEARER

The LM program may have ended in 1990, but the endurance racing flame wasn't completely squelched in Maranello. In 1986 Enzo Ferrari had hired a 21-year-old racer named Franco Meiners, and he would end up carrying the torch.

"I was introduced to Ferrari by friends of mine who were already working with Ferrari," Meiners said. "They wanted to build a factory-backed club that would be called Ferrari Club Italia. It would manage all the other clubs because there were hundreds of them in Italy and no coordination between them.

"After they accomplished this, they wanted to create a racing team for historic and production cars. Because I had been racing since I was a child, I knew these types of cars, the racing tracks, and the organizers. I was young and traveled a lot, so they asked me if I wanted to manage the team and organize this."

Meiners agreed and learned how to navigate the corridors of Maranello. That came in handy, for in 1990–1991 the speculative collector car bubble burst, which led to a decline in historic racing in Europe. Italian racing promoter Sergio Peroni recognized the trend, and spoke to Meiners about keeping some type of endurance racing alive.

"He approached me during a Historic car meeting at the Nürburgring and proposed a championship for supercars," Meiners recalled. "This meant cars like Porsche Turbos, Ferrari F40s, Jaguar 220s, Lotus Turbo Esprit, and smaller categories like the GT2 category and so on."

Both knew the key to getting the effort off the ground was a sustained Ferrari presence. "I promised him I would do it," Meiners explained, "but it was a bit of a battle with the factory. We were depending upon Ferrari production . . . for which type of car we could race. I was sure of what I was going to do and how to prepare the catalytic and noncatalytic F40s. I had to convince Mr. Fusaro of this, but he did not believe in the product and was putting his head in the sand. He was scared to death of not having the results we wanted to have—winning."

Whether that fear came from the lack of outright victories in IMSA or something else, Meiners never knew. But it didn't stop him. "I went ahead without authorization or support of the factory," he laughed. "The cars had to be close to production specifications so I had Adriano Nicodemi, a tuner in Padova, prep them and serve as pit crew. He wasn't at the level of Michelotto, but we didn't need that type of expense because not everything needed to be modified."

Peroni's new series was called the "Gran Criterium Supercar Italian GT Championship," and it was a throwback to the 1950s and early 1960s, where talented gentlemen racers could compete on a high level. The first race was held at Monza on March 15, 1992, and was won by the Ferrari Club Italia's F40 (chassis 89415).

"The (F40 LM) was simultaneously fabulous, and a little crazy . . . I ended up at right angles to the track more than once."

— Michel Ferte

"I still needed the support of the factory," Meiners said, "so I went back to the commercial and marketing people and said, 'first qualifying, pole position; first race, first place; second heat, first place. They didn't understand the F40's potential and were afraid that other manufacturers were ahead of us. So they replied, 'Bring us another race.'"

Which Meiners did two weeks later when chassis 89415 was victorious at Vallelunga. Another race was requested, and the results were the same. Amazingly, three straight victories didn't convince them, and Meiners blew up: "I said, 'That's enough!' and threw all the paperwork in the air."

The theatrics gave him the support desired, and F40s went on to win the remaining five races, besting a number of Porsche 964 Carrera RSs, several Lotus Esprit Turbos, an occasional Porsche 944 Turbo S, and even a Ferrari 348.

In 1993 the competition ratcheted up. The starting grids were larger and filled with Jaguar XJ220s, Porsche 911 Turbo S and Carrera RSRs, a large number of Mazda RX 7s, and Alfa Romeo SZs. Though F40s didn't sweep the season, they won eight of the ten races, and secured a second championship.

"These were basically production cars with catalytic converters," Meiners explained. "The factory was doing the work, and we made the gearboxes more reliable and modified the turbos to be less susceptible to breaking from the heat. The brakes were prone to cracking after racing, so we made them like the LMs and started to make the bodywork lighter."

THE F40 GT AND GTE

Competing against those factory cars would be a new F40 variant, the F40 GT. Engineered and built by Michelotto, the model had a 590-horsepower engine, lighter carbon-fiber coachwork, upgraded suspensions and brakes, and more.

The first F40 GT (chassis 74047) raced throughout 1992 and scored one overall victory and six seconds. The Padova-based firm made another six F40 GTs—chassis 79922, 80742, 84642, 88203, 90508, and 94362. The most successful was chassis 80742; it dominated the 1993 Supercar GT Championship by winning eight of the ten races in which it competed.

Though 94362 would win another Italian Championship in 1994, Meiners had his sights set on another target. "The life of a championship is always around three years, and then it starts going down," he reflected. "The costs become very high for the entries, so I did an agreement with Jurgen Barth to participate in a new championship he was building with Stephane Ratel and Patrick Peter."

The BPR (named after **B**arth, **P**eter, and **R**atel) series was launched in 1994 with eight races. Because Barth had come from Porsche and Ratel from France's small sports, GT manufacturer Venturi, the first few grids were filled mostly with cars from those marques. Meiners noted that "when we started in BPR in 1994, we ran two 348 LMs built by Michelotto and beat the Porsche 3.8 RSR."

That year, the Padvoa-based firm also built the most potent of all F40s. The F40 GTE (the "E" for Evoluzione) had a stiffer chassis, sequential gearbox, carbon brakes, improved aerodynamics, and an engine that was bumped to 3.6 liters for the 1995 season.

The first GTE was chassis 90001, and it debuted at the second BPR race at Jarama in Spain in 1994. It nabbed pole position, being 1.3 seconds faster than a Venturi 600 LM, but would retire on lap 62 with turbo issues. At 90001's fourth race (Vallelunga in July), the GTE garnered its third pole and took the checkered flag. At Spa it was second on the grid and came in second behind a Venturi.

That inaugural BPR season did not have a formal championship. In 1995 that changed, and some big supercar names—Jaguar, Bugatti, and McLaren, to name three—wandered onto the playing field. McLaren's F1 GTR became the car to beat, winning straight out of the box at the season opener in Jerez. Competing against them was a smattering of Ferraris—several F40s, the odd 355, and the Michelotto 348 LM.

⌃ In June 1994 F40 GTE chassis 90001 competed at Le Mans with completely different livery than that seen at Jarama. It qualified 14th on the grid but retired after 52 laps with electrical issues. It (along with two 348 GTCs and a 348 LM) was the first Ferrari to run at Le Mans since heavily modified Boxer chassis 35529 ran in 1984. *LAT*

Things became interesting at Monza when the F40 GTE (chassis 82404) competed for the first time that season. Meiners had vivid memories of the race, thanks in great part to McLaren engineer Gordon Murray: "He came over to take a look at the car and asked if I could show it to him. His team had two cars there, and the finishing was absolutely astonishing. After looking at the GTE he said, 'I've never seen a car with its chassis so badly welded.'

"I said, 'That may be, but this is a product from eight years ago. And while the chassis may be old, it's very effective.' I then offered to bet him that the F40 would be one and a half seconds faster on the track than his cars."

An astonished Murray replied, "No way!" and suggested they make a wager over qualifying. Out went the team's two best drivers—John Neilson in the GTR and Anders Olofsson in the GTE. "I had so much passion for the F40 model and what we had done, so I told Olofsson to really push it," Meiners laughed. "He held the F2 record at the Nürburgring and was great at setting up a car. The McLaren was much more modern than us, but it had a handicap in the heavier twelve-cylinder BMW engine."

Sure enough, Olofsson qualified on pole, beating Nielson by 1.6 seconds. Murray made good on the bet but had the last laugh. The McLaren won the race; the GTE came in third.

Which is how it went for the balance of the season. The GTE consistently out-qualified the McLaren, but the British car took the checkered flag. Only at Anderstorp in Sweden did an F40 take both the pole and checkered flag.

The following year's results were much the same, an F40 GTE nabbing the pole, the McLaren taking the victory. Once again, Anderstorp was favorable to Ferrari, with a GTE grabbing pole position and the victory. The F40s did not compete in 1997, as the BPR morphed into the FIA GT series.

FLAT OUT DOWN MULSANNE (FINALLY!)

A note must be made about the most fabled race of all. In 1994, Ferraris returned to the 24 Hours of Le Mans for the first time since 1984 when a lone F40 GTE (chassis 90001) was joined by a 348 LM and two 348 GTCs. Only one of the GTCs (chassis 97553) finished, placing eleventh OA. The GTE qualified fourteenth, and retired on the fifty-second lap with electrical problems.

In 1995, three F40 GTEs competed, qualifying sixth through eighth on the grid by running more than 11 seconds faster than 90001 did the prior year. Two GTEs finished, in twelfth (74045) and eighteenth (90001) positions. The following year three GTEs competed (88779, 75045, and 82404), but all were DNFs.

Joining the F40s in the last two years was Maranello's new purpose-built endurance racer, the 333 SP (see Chapter 6). Though that car would win fifty-six races and a good number of FIA and IMSA championships, it did not have the same results at Le Mans. In 1995 chassis 002 ran just seven laps before being felled by electrical problems. Two 333s ran in 1996, but neither finished after chassis 005 set the race's fastest lap. The highest placing occurred in 1997 when chassis 010 finished sixth overall.

But as will be seen, by then a much faster endurance racer was under development, one that was based on Ferrari's newest hypercar, the F50.

"I said to [Gordon Murray], 'While the chassis is old, it's very effective.' I then offered to bet him that the F40 would be one and a half seconds faster on the track than his cars."

— Franco Meiners

≋ At Le Mans in 1995 was F40 GTE chassis 74045, seen here during a pit stop. The car qualified 7th and finished 12th. 74045 returned to Le Mans the following year, where it was a DNF. *LAT*

≋ F50 production began in May 1995, with the first three cars going to the Sultan of Brunei. The first European car (chassis 103097) was produced on July 5, 1995, the first American car (chassis 103288) seven weeks after that. *LAT*

⌃ "The 1970s reflected a period of depression," Sergio Pininfarina noted. "After the March of 40,000, that changed. All of the 1980s were an exaggeration of flamboyance." Automotive design reflected that trend, and a darling of the decade was Ferrari's Testarossa, the car that starred in *Miami Vice*.

THE LATE 1980S WAS AN UNPRECEDENTED PERIOD in the sports, exotic, and collector car world. Throughout the decade many cars' values experienced a seemingly unstoppable rise, a stealthy trend seen only by enthusiasts, collectors, dealers, and the auction houses that catered to the hobby.

Then Black Monday hit on October 19, 1987. The Dow Jones Industrial Average plunged nearly 23 percent, wiping out more than $500 billion in wealth. NYSE chairman John Phelan described it as "the nearest thing to a financial meltdown I ever want to see."

Other markets fell further. By the end of October the Hong Kong Exchange had lost 45 percent of its value, Australia's nearly 42 percent, Spain 31 percent, and the UK more than 26 percent.

F50

Year(s) made: 1995–1997
Total number produced: 349
Hypercars: all
Drivetrain: 4,698cc DOHC V12, 513 hp @ 8,500 rpm; six-speed manual transmission
Weight: 3,080 lb (curb weight)
Price when new: $560,640

PERFORMANCE:

0–60 mph: 3.8 seconds
0–100 mph: 8.5 seconds
1/4 mile: 12.1 sec @ 123 mph
Top speed: 194 mph

Road tested in: *Car & Driver*, January 1997
Main competitors: McLaren F1, Lamborghini Diablo SE, Bugatti EB110 SS

As money flooded into cars, Ferrari—and everyone else—got caught up in the heat of the moment, and Maranello would produce nearly 1,000 more F40s than the 400 originally announced. That created a credibility problem for Ferrari, especially after the price bubble burst. *Michael Dregni*

The most obvious poster child of the excessive 1980s was Lamborghini's four-valve Countach. Like the Testarossa, it became known as "an investment vehicle," where demand exceeded supply, and prices escalated rapidly. Much of that upward spike was a spillover from speculative money pouring into the classic car market.

BOOM AND BUST

That frightening global decline, coupled with American tax law changes the previous year, precipitated a clamor for alternative assets. Investors soon discovered what auto enthusiasts had known for years: Certain cars showed stellar appreciation over time. Money flooded into the arena, with Ferrari values being the biggest beneficiaries. Prices went ballistic, and in October 1989 they hit their apogee when a Japanese buyer paid $14-plus million for a 250 GTO.

By then the speculative fever had spread far beyond classic cars. "We would routinely get double the sticker price for a new Testarossa," said Brandon Lawrence, the former general manager of a northern California Ferrari dealership. "We had consortiums of doctors and such that took their retirement plans and put the money into modern and older Ferraris. That's when I started to get nervous, for they didn't know a Dino from a Daytona, only that they had both."

That frenzy of viewing new cars as an investment goosed the hypercar market. It was the 1960s all over again, only on steroids. Speed and flash were hipper than ever, beaming into homes everywhere as Sonny Crockett tore up the streets on *Miami Vice* in a white Testarossa, hunting down the bad guys who were also piloting something fast and flashy. Manufacturers, established and not, focused their efforts on the performance market, wanting some of the glamour and prestige that came with being quicker than their competitors.

The crown jewel remained top speed. Stalwart contender Lamborghini gave its Countach four-valve heads, 455 horsepower, and a 185- to 190-mile-per-hour top end. Its successor the Diablo was quoted at 202 miles per hour, or 1 mile per hour faster than the F40. Elsewhere in Italy, former Lamborghini employee Claudio Zampolli had his prototype Cizeta V16T up and running, while Piero Rivolta was attempting to reboot Iso with a new Grifo. Mainstream manufacturers such as Japan's Subaru and Isuzu showed mega-fast mid-engine prototypes (the Jiotto Caspita and 4200R, respectively), Jaguar would produce its 213-mile-per-hour XJ220, and McLaren introduced its 230-plus-mile-per-hour F1. Ultra successful automotive dealer, importer, and gonzo car enthusiast Romano Artioli was the most ambitious of all, creating a stunning facility from scratch outside Modena to resuscitate the Bugatti marque and build the Nicola Materazzi-engineered 209-mile-per-hour EB110.

With the exotic car market so hot, and Enzo Ferrari no longer the gatekeeper (he died on August 14, 1988), any lid on F40 production was quickly swept aside. "Ferrari always spoke of making 400 cars," Materazzi recalled, "but [Ferrari CEO Giovanni] Razelli wanted 600. He was working for Fiat and wanted money, money, money. During the F40's introduction, he asked Ferrari, 'How many cars do I have to present?' Ferrari's hand was hidden by the table, so he replied '400' by showing four fingers."

The number didn't stay there. At the model's formal launch in June 1988 the magazine reports noted "more than 900" were to be built, and the original price of $200,000 had also grown. It was listed as $260,000 in *Autoweek*'s June 13 article, ballooning to $400,000 when F40s finally hit U.S. shores. Prices were approaching $1 million on the secondary market, so Ferrari kept production going into 1992, ultimately building 1,311 cars.

By 1992, though, the collector car–exotic car bubble had burst. A sudden, nasty global recession hit in 1990, sending real estate prices crashing (especially in Japan, which had fueled much of the speculative automotive bubble). America's loosely regulated savings and loan industry cratered, resulting in a several hundred billion dollar government bailout. Plus there was the brief Gulf War in 1991 and a financial system awash in leveraged induced-wealth and liquidity found itself deleveraging at an astounding rate.

"The market went into a freefall," Lawrence explained. "I remember my boss telling me, 'Ferraris have leprosy; nobody wants them.' In 1990 he took a half-million-dollar hit in a number of months on a car he bought."

Others saw even larger losses. That $14 million GTO would sell for $3 million several years later, and demand for Testarossas, Countaches, F40s, and all other hypercars evaporated. Prices spiraled downward. Jaguar was hit with lawsuits regarding its XK220, Artioli's Bugatti adventure folded in the mid-1990s, and McLaren's planned 300-unit production run for its $1 million F1 struggled to get to 107 by 1998. F40 prices dropped to $250,000 and less, that reduction no doubt fueled by over-production.

From all of this Maranello learned a lesson: It would never over-produce a hypercar again.

CHANGING THE TARGET

Ferrari's first hypercar to be built under its "one less than perceived market demand" philosophy was the F50, a car conceived during the height of the performance feeding frenzy.

In the autumn of 1989, with F40s coming off the lines as quickly as they could be assembled, Ferrari vice chairman Piero Ferrari started to ponder what the next hypercar should bring to the table. He often drove an F40 to the office, but rather than focusing on headline-grabbing numbers such as top speed, g forces, and quarter-mile times, he chose a courageous and much more elusive criteria—a driving experience not found anywhere else.

"Clients would come to Maranello and ask, 'Why can't Ferrari make something close to a Formula 1 car we can drive on the road?'" he told Peter Robinson in the *Autocar* booklet *F50 The Ultimate Ferrari.* Having F1 as the basis for the next "ultimate" was logical, as Ferrari's front-line endurance racers—machines that often served as the engineering test bed and starting point for the original hypercars—hadn't been the company's focus for two decades.

≈ For more than two decades, the man who altered the hypercar game has gone unrecognized. In the late 1980s, when F40 production was going great guns, Piero Ferrari contemplated something that up until then had been unthinkable—no longer chasing top speed. He's seen here in his office in 1997.

« What makes Piero Ferrari's "top speed is no longer important" edict all the more remarkable was long after it was made, McLaren's 230-plus-mile-per-hour F1 broke cover. That recalibrated the top speed bar by a substantial margin, but Ferrari didn't budge—205 miles per hour was fast enough, a decision that would completely alter the hypercar game several years later. *LAT*

⌃ As one who grew up in the midst of Ferrari history, Piero Ferrari has an uncanny ability to see it from an outsider's point of view. Here he is in Maranello's Classiche Department, where historic Ferraris are restored and certified.

Piero Ferrari is the last of Maranello's old guard. The second of Enzo Ferrari's two sons, he was born on May 22, 1945, to Enzo's longtime paramour Lina Lardi. Though his childhood interests were "the sea, sports, powerboats, and, in the winter, the mountains," what really captivated him were the engines he saw being repaired and rebuilt by a motorcycle mechanic whose shop was in front of his house.

Ferrari entered college in 1964, studying economics because "my father pushed me in that direction." Any thoughts of getting a degree were cut short when he started working in 1965 in Maranello. Enzo's mother had told Ferrari shortly before her death that Piero's wish was to work in the factory.

"My world changed the first time we went there," he explained with a smile. "We went out to a restaurant for dinner, and after that he took me over to the company. It was night, and everything was very quiet. He didn't say a word as we walked through the buildings, seeing the racing department and the production lines. I'm sure my eyes were very big, looking at everything."

He began working alongside Federico Giberti, an Enzo confidant from the Scuderia Ferrari days. "Giberti was responsible for the purchasing department and introduced me to the most important suppliers. I began looking at the problem of components and manufacturing. My first job was when my father decided to produce the Dino Sport [206 S]. The prototype was made in the racing department, and I organized all the components to produce them.

"The cars were made in the production department, not the racing department. At that time everything was much smaller, much easier, less complex. I had a copy of the drawings the racing department gave to the purchasing departments and would make all the orders to the suppliers. I also organized the production of the cars."

After the 206 S Piero moved to the racing department, staying there until his father's passing in 1988: "I worked with suppliers and planned the orders of components. I also watched over the people, employing the mechanics, and organizing the racing team.

"My first real involvement [in everything] was in the Lauda period, when Montezemolo came to Ferrari, and Forghieri was the technical director. . . . From the mid-1970s on, my father was using me like an interface with a lot of problems. He was dealing through me with many people inside the factory and was using me—I was maybe a tool of my father—because I was available even in the evening, when we were at home. He was living on the first floor, I was living on the third, and after dinner he was calling, 'Piero, come down, come down.'

"Every evening he had a problem. 'Remember tomorrow,' he would say. It was very tough to work with my father. He was not an easy man because he was asking to the employees, not just the top employees, too much. And to me, even something more."

Piero's duties branched into following the regulations and working with Marco Piccinini in dealing with Bernie Eccelstone. By then, Enzo had become "more than just dad" to him: "When I started traveling, especially to America and Japan, I realized that the Ferrari name was so famous. Probably my father didn't know he was so well-known around the world."

Since 1988, Piero has been a company vice president, and a 10 percent shareholder. Speaking about Maranello's competition heritage, "In Ferrari's early years endurance racing was more popular than Formula 1," he explained. "It was really . . . what built up the Ferrari myth in that age. Before the 1970s . . . [the season] was maybe nine or ten grand prix, plus six endurance races. It was like today where we do eighteen races in Formula 1, [but] to compete [now] you have to make a choice. Audi, they have made a very strong commitment to Le Mans and did very well lobbying to have the diesel, to make it winning Le Mans. Okay, they did a good job, but this proves you have to be committed to one kind of competition and not both."

On what he enjoys most at Ferrari, "to be involved in projects. When we have to make a new car or a new engine—a new something—I'm happy to do it. That is my mentality."

He's had key roles in all of Ferrari's modern-era hypercars, but the three machines that are considered "his" are the F50, F50 GT, and 333 SP. About the 333, Ferrari's last true purpose-built endurance racer, "The idea came from my relationship with Gianpiero Moretti at MOMO. A good friend for over twenty years, he dreamt of finishing his racing career driving a Ferrari. He came to me and said IMSA was making new rules and a new formula, and the ideal car was a V-12 derived from the F50.

"I went to Montezemolo and said, 'We have a chance to make a competitive sports car.' . . . He accepted the proposal, and we made the car with the help of a few F1 engineers." Though the business plan was built around a production run of ten cars, forty were delivered to customers.

What's surprising to learn is Piero has a small company that's a NASCAR subcontractor. "It's an engineering firm that I started in 1998. I try to spend [some] time by myself and my engineers in some fields or areas of technology to be like a hub, to exchange experience from Formula 1 to NASCAR and vice versa.

"I've been to the Daytona 500, which is really quite the motor racing event of the year in the U.S. . . . What I really like [is] you see the cars, all tubular frames, bodywork with metal sheets and not carbon fiber. It looks like quite old technology but behind it there is a lot of experience and every detail is developed to perfection, as they have not much space to do new technology. But they know every screw of the car."

Now ending his fifth decade in Maranello, Piero remains the company's largest noncorporate shareholder. Through it all, he has stayed extremely humble, and well-grounded—most remarkable, considering his name is on the building. "He's been very true to himself," a friend noted. "He's never tried to be someone or something that he is not."

≈ To understand the depth of Piero's roots in Maranello, his first job at Ferrari came in 1965, when he was put in charge of organizing all the parts to produce the 206 S (right, chassis 014 at Le Mans in 1966 with 275 GTB/C chassis 9015). *Jean Charles Martha collection at the Revs Institute for Automotive Research*

≈ Piero was given more duties in the 1970s and 1980s, working in the F1 department and acting as a conduit throughout the company for his father. This is at the 1981 annual luncheon, where he was involved in the annual award ceremonies.

If any photograph shows the growth of the Ferrari firm, it's this one. Behind Piero lurks the small city that Ferrari's factory grounds have become—it's now so large that bicycles are needed to get from one side of the facility to the other in any sort of punctual manner.

Piero's reply to those customers highlighted the pivot to come: "Maybe 330 kilometers per hour (206 miles per hour) is fast enough. Perhaps we should make something closer to Formula 1 technology. . . . "

"People said the F40 had a fantastic engine," he told Robinson, "but it was still an eight-cylinder. Some of the old clients were dreaming only of a V-12, the classic Ferrari engine. So this was our first choice. . . . Our second idea was to develop a real carbon-fiber chassis like a Formula 1 car. To combine the two was the beginning of the idea for the F50."

What made the target especially bold was Maranello's multiyear bad luck streak in F1. From 1975 to 1983, the Scuderia won nine titles—six constructors' championships and three drivers' titles. From 1984 through 1989 it won a total of nine *grand prix*.

In retrospect Piero's timing couldn't have been better, because the team's fortunes were about to take a decided upswing. In 1989 F1 launched a new naturally aspirated 3.5-liter engine formula, effectively ending the sport's turbo era. That season Ferrari ran a 3,498cc V-12 and won three races. The following year the engine gained just a single cc but boasted an additional 80 to 100 horsepower. Ferrari had also signed three-time world champion Alain Prost to partner with fan favorite Nigel Mansell. Prost would score five victories while narrowly missing the drivers' title, while Mansell managed a single victory.

In early 1990 Piero's F1-racer-for-the-street concept took hold during visits to

≈ A technology Ferrari considered for the F40 replacement that wasn't quite ready for prime time was a hydraulically actuated, Formula 1–style paddleshift transmission. That didn't mean it wasn't being tried, though—it was used in the secretive Ferrari FX (shown is chassis 103396) made by Pininfarina for the Brunei royal family in the mid-1990s.

the racing department, a separate facility about a mile away from the factory. During brainstorming sessions with some of the department's key personnel, "we began to put together the various pieces of the car," he told Robinson. "We decided to develop the then current 1990 3.5-liter F1 engine from the 641/2 race car. It was very compact, very short, whereas our other (road car) V-12s . . . would be very difficult to fit in the rear without making the wheelbase too long. We didn't want a big car with a big engine; our desire was to have a compact engine and a low center of gravity."

By the middle of the year the team was using a wooden model to simulate the car's seating position, wheelbase, and track. Once the general dimensions were determined, and the car's mechanical parameters had become fairly concrete, a call was placed to Pininfarina to discuss design.

DESIGN HICCUPS

Overseeing the effort was Lorenzo Ramaciotti. Then forty-two years old, his career in Pininfarina started in January 1973. "I wanted to be a race car engineer coming out of college," he said. "My dream was to be the Italian Colin Chapman, but I think it's important to grab a chance when it comes, so I grabbed the chance from Pininfarina."

He ended up with the perfect instructor, working under Leonardo Fioravanti in several capacities—designer, assistant, and shop manager. In those formative years he

≈ When the Mythos was wowing the show circuit, the man leading the Pininfarina design team was Lorenzo Ramaciotti (seen here with Sergio Pininfarina). "It is one of my favorites that I was involved with," Ramaciotti said. His first car as design director, "it assured me that I was able to produce concepts at the level Pininfarina needed, and was expecting." *Pininfarina archives*

The FX was based on Ferrari's top line production model at the time, the F512M. The custom interior used the finest materials and featured red and green paddles behind the steering wheel for shifting. The gear changes felt very tentative and mechanical compared to today's instantaneous shifts.

discovered his real strength was in "building teams, having them work in a cohesive way, where you find a direction and have them focus on that. I like to have a comprehensive view, as I see a project as bringing things together."

In January 1988 he succeeded Fioravanti as director of Pininfarina's Studi e Ricerche on the outskirts of Turin. One of his first projects was the Mythos, a visually cohesive and powerful-looking barchetta done on Ferrari Testarossa mechanicals. The prototype debuted at the Tokyo Motor Show in 1989 and was so stunning that it won a number of major design awards, including that year's most innovative concept.

Less than two years later Ferrari was briefing him on their next hypercar, codenamed the F130. "In general," Ramaciotti recalled, "their idea was to make a successor to the F40. My team of designers was more or less the team that worked on the Mythos, and we started with a very flamboyant design in terms of proportions. The car was much more cab forward with a long engine in the back, as we tried to stress the length of the engine. The 288 and F40 were turbocharged eight cylinders while this was a naturally aspirated twelve-cylinder."

CAD was used to help finalize the design. A 1:3 scale model was made, and Ferrari authorized wind tunnel testing. A flat undertray with twin diffusers optimized the aerodynamics. High downforce came from a large wing at the rear.

But unlike the 288 GTO and F40, the F50's design process was muddled by several factors, including indecision over whether the car should be open or closed. "We were

Whereas the FX used the production F512M's flat-12, the F50 had a 4698cc V-12 powerplant that was derived from 1990's 641/2 F1 car. *Ferrari S.p.A.*

really confused," Ramaciotti recalled. "It was done in a transition time, between the death of Enzo Ferrari and the hiring of Montezemolo, when the company was steered by Piero Ferrari and Piero Fusaro.

"The F40 was so successful, and such a seminal vehicle with the visual impact of the wing and that front, that we wanted to do something different. At the same time, they were scared of doing something that was too far from the F40. They feared if it was too far it would not have the same visual impact and continuity so there was kind of a struggle where we had to do it different, but not too much.

"We started with a berlinetta then tried to do a barchetta, and finally did something that was a compromise between the two. They were a little scared of the heritage they had to bring from the F40 so they tried, and we also tried, to embody into the car everything, and the opposite of everything."

The final 1:1 styling model was completed in June 1991 and was nearly identical to the production car. In an effort to quell the berlinetta versus barchetta debate, Ramaciotti had the model's roof painted black so the roofline would mimic the appearance of an open car. In the end, it was decided that the F50 would be both open and closed.

"They tried to make a car that would please everybody," Ramaciotti explained. "It was a coupe that could also be a barchetta, a kind of wish of making everybody happy. . . . It was not easy, like we were working with the handbrake on."

A BOLD TECHNOLOGICAL LEAP

The first development prototype was running in 1991 and underwent an unusually lengthy engineering process. "The car marked a real change," Ramaciotti noted. "There was a radical difference between it and the F40 because the F40 had the same structure as the GTO and in that period, the structures at Ferrari were very crude. It was steel tubing, and the lightweight materials, or I would say the most sophisticated resins, were not a major part of the structure. They were in the body and in some reinforcement here and there.

"The F50 really changed the picture as it was 100 percent carbon fiber. So the step of development between the F40 and the F50 was even bolder because it was really a technical direction and layout of a Formula 1 car, [where] the engine was rigidly bolted to the structure and acted as a stressed element."

The F50's avant-garde suspension design fell on the shoulders of Ferrari engineer Carlo Della Casa. Each corner featured double wishbones, coil springs over electronically controlled Bilstein shocks, anti-roll bars, and all-metal ball joints. The uprights were made of titanium, and the coil-over-shocks used F1-inspired pushrods. In back the wishbones were mounted on the engine and were extra-long to exert more control over the shocks and give the car a more supple ride. The ECU controlled each shock separately and worked off of the car's speed, steering wheel angle, and the body's vertical and longitudinal acceleration.

Carbon fiber was considered for the large, unassisted 335mm ventilated cast-iron discs but not used because the material took too long to warm up. The wheels were made of magnesium, 8.5x18 inches up front, and 13x18 inches in back. Steering was unassisted rack and pinion to keep feedback as pure as possible.

Prior to the F50, all of Ferrari's road and race V-12 engines had a 60-degree vee. For the F50, the vee would be 65 degrees, as seen in the Formula 1 powerplant. "I [wanted to] package the [F1] unit in the best possible way," former Ferrari chief engineer John Barnard told Alan Henry in *Ferrari, the Battle for Revival*. "We originally thought of a 60-degree layout, but opened it out another 5 degrees in order to give space for more ancillaries within the vee."

That 4698cc V-12 had an 85mm bore, 69mm stroke, and aluminum heads. In a departure from typical Ferrari practice, the block was made of cast iron rather than aluminum since the engine would act as a stressed member of the chassis. The heads

⌃ The F50 really changed the picture as it was "100% carbon fiber," Pininfarina's Lorenzo Ramaciotti noted. The central structure of the car was its lightweight, open roof tub. *Ferrari S.p.A.*

The cover is pulled back at the F50's 1995 Geneva Show introduction. Piero Ferrari has his back to the camera, while facing it are Sergio Pininfarina and three-time F1 champ and former Ferrari driver Niki Lauda, who assisted in the car's development. This reveal marked the first time in ages that Ferrari's performance flagship had a mixed reception. *Pininfarina archives*

featured DOHC and five valves per cylinder, three for inlet, two for exhaust. Oil came from a separate dry-sump tank with three pick-up points. The Bosch Motronic 2.7 ECU had sequential fuel injection and an ignition system without distributors.

"The challenge was to take the 3.5-liter engine and increase the stroke to make the maximum possible capacity inside that block," Ferrari told Robinson. "I remember that for the first engine tests on the dyno we used the same old heads from Formula 1, so to tell you the engine is born as a development of Formula 1 technology is true. At the time it was 4.5 liters . . . because it was using existing parts from the racing engine."

Final power output was 513–520 horsepower (depending upon source quoted) at 8,500 rpm, and 347 lb-ft of torque at 6,500 rpm.

In Piero's original "F1 car for the road" concept, the transmission would be electro-hydraulic, along the lines of the revolutionary system Ferrari introduced in Formula 1 in 1989. An experimental paddle-shift semi-automatic was used in the FX (a special based on an F512 M that was specially built for the Sultan of Brunei), but the system would not be seen in normal road-going Ferraris until the late-1998 355 F1. "We didn't have enough time to develop a semi-automatic gearbox," Piero said. "With the F1 cars you always have your fingers on the levers," Ferrari engineer Tino Carniglia explained further, "but that's not possible on the F50, as the wheel turns through a much bigger angle."

The F50 thus used a six-speed that sat behind the engine, housed in an aluminum casting. The long gear linkage followed F1 practices by having a rod mechanism comprised of four sections, four universal joints, and two sliding bearings.

When testing commenced at Fiorano, a difficult challenge popped up. By bolting the engine and gearbox directly onto the chassis, with the powertrain acting as the sole support for the rear suspension, every sound and vibration (no matter how minute) was transmitted directly into the cabin.

The production F50 got 65 pounds of a resin-like material in the inner bulkhead and under the chassis to reduce the vibration and noise, but "it is still *the* question mark about the car," Ferrari said to Robinson at the time production started. "I was concerned about the noise problems [but] the intention was to make a car that you could use, even on the road, to generate something very close to the feeling that you have when driving a race car. [It] is very different to the one you get with road cars. A stiff chassis, aerodynamic downforce, and very stiff suspension with pushrods help take handling to the highest level."

Test driver Dario Benuzzi put more than 15,000 miles on the prototypes and was quite pleased with the results. "To understand the character of the F50," he told *Autocar's* Steve Cropley, "you have to use all the power and get it cornering near the limit—and that takes a pretty good driver. But when you do it, you feel the responses of a Formula 1 car and hear the music. It's closer to a single-seater than any road car we've ever built.

"Of course, the F50 is not as quick as the single-seater. Nothing is. And maybe it is not quite as agile. But the speed of the steering response, the feel of the brakes and clutch—and especially the instant answer you get from the engine when you squeeze the throttle—they are really very similar to a grand prix machine."

Lending much input into the development process was Niki Lauda. "I drove various prototypes for almost two years before it was unveiled and was called in to assess every key modification," the former Ferrari driver and three-time F1 champion told Alan Henry in *F50 The Ultimate Ferrari*. "Driving the F50 round our Fiorano test track, well, the F40 just doesn't stand comparison, from a performance point of view. [The F50] has completely neutral handling . . . brakes like you cannot believe, and has a brand new gearbox which is . . . not a typical Ferrari-like change.

"The big challenge for the engine department was to balance the supreme F1 power with sufficient drivability to make it acceptable for the roads. From 8,000 rpm it comes on with such a surge that you just won't believe it."

AN UNEASY RECEPTION . . .

Spy shots of the undisguised development F50s were seen in 1994, and the cover came off the production car at the 1995 Geneva Auto Show, where the reception was decidedly mixed. Yes, it was Ferrari's newest hypercar, but its quoted top speed of 202 miles per hour was just a single mile per hour faster than the F40 and a good pace off the McLaren and others. Plus acceleration times didn't move the bar, as had happened when the F40 came out.

Beautifully summing up everyone's reservations was *CAR's* April 1995 headline. "Ferrari F50," it read, "quicker than an F40. Just."

An even bigger question was how all this news would play out in an exotic car market still reeling from the recession-induced demand meltdown. "It is quite important to the health of Ferrari's finances that the F50 find the required number of buyers," *Autocar's* Steve Cropley observed. "If things were to go wrong it would hurt grievously, just as it has stung the creators of other expensive cars that have fallen on hard times, notably the Jaguar XJ220 and Bugatti EB110."

And there was the lingering sting from the F40's huge production run. "The people who buy early F50s will do it on the understanding that the batch will be strictly limited to 349 cars," Ferrari sales and marketing director Michele Scannavini told Cropley. "We'd be breaking faith with those people if we built more. Each owner will get a certificate, signed by our president, which has something like '200th out of 349' on it.

"The one event which may swell the total number beyond 349—by no more than 20 cars—would be a decision to build race spec F50s." Which, as will be seen, very nearly happened.

continued on page 171

≫ At Geneva's second day are Sergio Pininfarina, Piero Ferrari, and Ferrari's CEO Luca Cordero di Montezemolo, who had been at the helm since late 1991. Over the next two decades, Montezemolo would revolutionize Ferrari in many ways. *Pininfarina archives*

» A brilliant image at Fiorano as Ferrari test driver Dario Benuzzi puts what is likely F50 chassis 99999 through its paces. The F50 was more than three seconds faster around the track than its F40 predecessor. *LAT*

Two pan shots show the F50's profile. While the car lacks the fluidity of the 288 GTO and F40, neither comes close to sounding like an F50 under full throttle. Especially in guise most every owner keeps the car—with the top off. *John Lamm* (above), *LAT* (left)

≽ On the production line, the business end of the F50's drivetrain shows the rear-mounted six-speed transmission and how it has the pick-up points for the rear suspension. *LAT*

The center section of the chassis was a tub made entirely of Cytec Aerospace carbon fiber. Here are the powertrain, windscreen surround, and front steel frame that houses the radiator and cooling fans. *John Lamm*

When F50 production began, there was real concern in and outside Ferrari that all 349 cars would not find owners. The economic environment was quite different from the yeasty late 1980s, plus some clients were reluctant to commit to early cars because of the continual escalation in F40 production. *John Lamm* (above), *LAT* (below)

continued from page 163

. . . IS QUICKLY SWEPT ASIDE

All the pre-sales hand-wringing was quickly forgotten once owners and the press got behind the wheel. They discovered the F50 was indeed different—and incredibly memorable.

The first journalist to try it on public roads was likely Paul Horrell in *CAR*'s March 1996 cover story. "[The F50] is stable and content at speed," he wrote about a number of blasts to 160 miles per hour on the empty roads outside of Dubai, "and I know it can stop. The brakes are colossal, non-servoed, and breathtakingly able. The pedal asks for a hard push [and] the more you ask, the more it delivers. This car loves big speed, but more than that it loves changes: accelerating, braking, cornering. For these it was undoubtedly made."

Of special note was the transmission and engine. About the former, he felt "It's soon clear that this, Ferrari's most frenetic car, has the company's easiest ever gear change, light and sweet in its travel yet still fast and micrometrically precise."

And though the V-12 lacked real oomph below 4,500 rpm, "in second gear, the moment you open the taps should be the same moment you prepare to change up. The interval between the two events is negligible. All you register is a moment of calamitous, delicious, exhaust-explosion music, and a rabid bolt of g-force. Grab third [and] again you barely mark the passage of time before fourth is called for."

When *CAR*'s December 1997 issue pitted an F50 against nineteen other cars on the 1.84-mile Castle Combe racetrack to find the year's sweetest-handling car, the Ferrari easily beat them all. "So much power, so much sound, such huge cornering forces," their editorial team concluded. "Yet all so friendly, so easily modulated. Has any other car this monstrous ever handled so beautifully before? We think not."

continued on page 177

≈ The F50 interior reflected Ferrari's decades-long experience of making barchettas—those open-air two seaters where everything unnecessary to extracting performance is deleted.

« All the concern at the F50's launch of cars going unsold proved unfounded, a main reason being the outstanding road manners. "Has any other car this monstrous ever handled so beautifully before?" *CAR* queried in December 1997. "We think not."

"[Ferrari's management was] a little scared of the heritage they had to bring from the F40 so they tried, and we also tried, to embody into the car everything, and the opposite of everything."

— Lorenzo Ramaciotti

F50 shows the classic Ferrari and Pininfarina curves, but the form doesn't hang together and have the coherence like the best designs. The reason is simple—the design team was confused, for their marching orders from Maranello changed from a closed car to open, and then finally to both.

≈ The F50 was the last of the Ferrari hypercars to use a manual transmission. Interestingly, when talking with collectors who own a number of the hypercars, their favorite to drive is most often the F50.

» Engine compartment eye candy doesn't get much better than this. A true visual feast, everything is there for a reason, not just to look pretty.

"From 8,000 rpm on it comes on with such a surge that you won't believe it."

— Niki Lauda

continued from page 171

The F50 wasn't bad in a straight line, either. Peter Egan's test in *Road & Track*'s January 1997 issue at Fiorano saw the Ferrari hit 60 in 3.6 seconds, 100 in 8.0, and the quarter in 12.1 at 124.5 miles per hour. "More than anything," he observed, "the F50 simply feels like a racing car. It's as pleasant and predictable to drive as my Reynard F2000 car, but with the bonus of an extra 400 bhp, fat tires, and a second seat. And air conditioning!"

Car & Driver's test at the Transportation Research Center's 7.5-mile oval in Ohio recorded much the same times. Successful entrepreneur, racecar driver, and Scandia team owner Andy Evans drove his personal F50 (car 003) for the test, with the odometer reading just 31 miles. He saw 60 in 3.8 seconds, 100 in 8.5, and the quarter in 12.1 at 123 miles per hour. Top speed in sixth gear was 194 miles per hour at 8,640 rpm or 8 miles per hour below the quoted maximum.

But the most telling performance number was one never found in road tests. In 1984, the 288 GTO became the standard bearer around Fiorano with a lap time of 1:36.0. The F40 turned a 1:29.6. The F50 dropped it to 1:27.0.

Unbeknownst at the time, that 1:27 marked the end of an era. Behind the factory walls in Maranello, a quiet revolution was taking place. Up till then, pure driving skill coupled with a car's mechanical components and capabilities, set up, and tires determined how fast it could go. By the end of the decade two new elements—electronics and software—entered the speed game, and Ferrari would master them better than anyone else.

But that didn't mean analog cars were said and done in Maranello.

OPPOSITE:

Sergio Pininfarina remembered admirers often asking to see the engine compartment when his company would exhibit Ferraris at auto shows. In response, they began putting clear covers on the engine compartments.

⌃ The mesh at the rear not only helped with engine compartment cooling, but it also gave a glimpse of the F1-inspired engine and suspension, while the giant wing looks every part of a Group C racer.

Endurance Racing

WITH AN F50 HEART

The first act of the endurance racing renaissance that swept through Maranello was the F40 LM in 1989–1990. From 1992 through 1994, F40s dominated Italy's racing scene and continued competing internationally in 1995 and 1996 with the Michelotto-engineered and built F40 GT and GTE.

Then came Ferrari's ambitious 333 SP, a car with an F50 link and Maranello's first purpose-built road racer in nearly twenty-five years. Back in the 1990s race driver Gianpiero Moretti was CEO of MOMO, the world's largest maker of aftermarket steering wheels and other accessories. He had raced a Ferrari 512 S at Daytona in 1970 and wanted to finish his career in a Ferrari.

In 1993, "[Moretti] came to me and said IMSA was making new rules and a new formula," Piero Ferrari recalled. Ferrari warmly welcomed the pitch but told his friend "the problem will be convincing others in management that this is a good idea. The effort has to be successful as well as make economic sense."

A plan was put together, and Piero approached company CEO Luca di Montezemolo. "We investigated and found it was possible to make ten cars," Ferrari said. "I told Montezemolo maybe we make some money, but for sure we wouldn't lose. He accepted this."

Also backing the project was Ferrari North America CEO Gian Luigi Buittoni. Sales had seriously slumped in Ferrari's largest market, and a return to a high-profile racing series such as IMSA held much promotional value.

⌃ Ferrari's first purpose-built endurance racer that actually competed since the early 1970s was the 333 SP, seen here racing at Sears Point in California in July 1996. Forty would be made by Dallara and Michelotto, and the car would score nearly sixty victories. Only an overall win at Le Mans remained elusive.

"It was possible to do the 333 in house, but they were always so busy with F1," Ferrari noted. "There had to be a group dedicated to designing the new car, so we decided to go to Dallara because he had the potential to do it in very few months."

The talented engineer worked with key Ferrari personnel to create the 333 SP. It used a carbon-fiber tub, independent pushrod suspension, five-speed sequential gearbox, and a body that was largely developed in Dallara's wind tunnel. The engine was derived from the F50's powerplant, reduced to 3,997cc. In final tune Ferrari quoted 650 horsepower at 11,000 rpm.

The model debuted at Road Atlanta in 1994 and took the top two spots. Over the next several years it was raced by a number of top privateers and would score another fifty-five wins and numerous championships; this included victories at famed tracks such as Daytona, Sebring, Watkins Glen, and the Nürburgring. Taking a number of those checkered flags was Gianpiero Moretti.

From 1994 to 2001, Dallara and then Michelotto made 40 333 SPs. "I was surprised at the car's longevity," Ferrari commented about the model's continued success. "After one year a racing car is old, so to see it still competitive at a championship level for that long made me happy because the 333 was a good project."

« The man mostly responsible for the 333 SP's birth was Piero Ferrari's friend, team owner, and longtime racer Gianpiero Moretti. He wanted to finish his driving career in a Ferrari, thanks to new IMSA rules that would allow the 333's powerplant to be derived from the F50's V-12.

333 SP

Year(s) made: 1994–2001
Total number produced: 40
Hypercars: none; all competition cars
Drivetrain: 3,997cc DOHC V-12, 650 hp @ 11,000 rpm; five-speed sequential gearbox
Weight: 1,936 lb
Price when new: $950,000

PERFORMANCE:

0–60 mph (mfr): 3.3 seconds
0–100 mph: n/a
1/4 mile: n/a
Top speed (mfr): 228 mph

Road tested by: n/a
Main competitors: McLaren F1 GTR, Porsche 911 GT1, Riley & Scott Ford Mk III, Riley & Scott Oldsmobile Mk III, Ferrari F40 GTE

ABOVE and OPPOSITE:

The 333 was the fastest non-F1 car around
Fiorano until the F50 GT came along. This
remarkable Ferrari was at the center of Piero
Ferrari's desire to return to the winner's circle
at Le Mans. This is F50 GT chassis 001, the only
one of the three that was made in the *Gestione
Sportiva* (Ferrari's F1 department).

F50 GT: FASTEST OF ALL

During testing, the 333 SP turned a 1:11 lap at Fiorano. As the car campaigned in the
1995 and 1996 seasons, an even faster endurance racer was undergoing design and
development. That car was what Piero Ferrari called the F50 *Competizione*, and it became
known as the F50 GT.

An extensively modified F50, "It was a car that was supposed to race in the 1997 FIA
World Championship," Franco Meiners said. "When the BPR stopped, it was because
Eccelstone took over the championship. He practically stole it from Patrick Peter so there
was a fight, which Patrick Peter won. This was called FIA World Championship GT, and
everybody was producing a car to participate in that series."

The F50 GT was an F50 with the dial turned up to 15. General mechanical specifications
remained the same—carbon-fiber monocoque, independent pushrod suspension front
and rear, rack and pinion steering—and then the fun began. The body was once again
made of carbon fiber, but modified with larger air channels and a central NACA duct on
the front clip. An intake on top of the cabin drew air into the engine compartment, while
in back a sloping fixed rear roofline gave the car a classic "berlinetta" look. The large rear
wing had central struts, and underneath the rear end was a diffuser to channel airflow.

The DOHC, five-valve-per-cylinder V-12's capacity remained at 4,698cc, but
compression was upped from 11.3 to 11.9:1. The Bosch Motronic injection and
ignition were replaced with more advanced Magneti Marelli systems. These and other
modifications saw horsepower increase from 520 in the F50 to a Ferrari quoted 750 at

10,500 rpm, and torque jump 43 lb-ft to 390 at 7,500 rpm. The transmission was now an X-trac sequential six-speed, and the multi-plate clutch was made of carbon fiber.

As much mass was pulled out as possible, the dry weight dropping from a Ferrari-quoted 1,230 kilograms (2,706 pounds) on the production F50 to 860 kilograms (1,892 pounds) for the F50 GT. Acceleration was listed at 0 to 62 miles per hour in 2.9 seconds, and top speed at 376 kilometers per hour (233 miles per hour).

F50 GT

Year(s) made: 1996–1997
Total number produced: 3
Hypercars: none, all competition cars
Drivetrain: 4,698cc DOHC V-12, 750 hp @ 10,500 rpm; six-speed
  sequential gearbox
Weight: 1,892 lb
Price when new: n/a

PERFORMANCE:
  0–60 mph: 2.9 seconds
  0–100 mph: n/a
  1/4 mile: n/a
  Top speed: 233 mph
Road tested by: Some very lucky drivers, only on track
Main competitors: McLaren F1 GTR, Porsche GT1, Mercedes Bens
  CLK GTR

» The interior of F50 GT 001. Two decades on, the F50 GT remains the fastest non-F1 car around Fiorano, and by a good margin. Depending upon the source quoted, it was one and a half to three seconds faster than the next fastest non-F1 car, the championship winning 333 SP.

"Ferrari decided to build the F50 GT, with Dallara making the chassis, and the engine done by Claudio Umbarti," Meiners explained. "It was to be sold to top level privateer teams. The first one [chassis 001] was built in a period where I left the factory and went to work with Benetton. I had an agreement to buy the cars, and race them as Benetton with Benetton colors. I ordered the car and everything and followed the evolution of it very closely.

"Dallara was doing the development, and Nicola Larini was the test driver in Fiorano. There really weren't any problems except an oil tank needed to be made a bit higher in the back. Everything else was spot on; it was a great car and would have raced against the Porsche GT1, McLaren, Panoz, and all the others."

As development testing continued, it became clear Maranello had one seriously quick and very tossable machine. One insider recalled it being 3.5 seconds quicker around Fiorano than the 333 SP, while Meiners said, "It was a second and a half faster than the 333 and possibly more. It was just great, an absolutely fantastic car."

Then, in the fall of 1996 as everyone geared up to make a serious run at an overall win at Le Mans, the effort was unexpectedly canceled. Rumors swirled in Ferrari circles, with various theories bandied about. One centered on the F1 team being in a tremendous slump since 1990, winning only two GPs from 1991 through 1995. Michael Schumacher had joined in 1996 and won three races, so would a potential Le Mans win overshadow the rebuilding of the F1 team? At least one insider claimed this was a real source of conflict in the management hallways in Maranello, so the program was canceled.

Others said Eccelstone didn't want Ferrari competing in Patrick Peter's series. Another camp stated Ferrari didn't want to take on Porsche, Mercedes, and other larger manufacturers.

Whatever occurred behind the scenes, it was clear Piero Ferrari was still smarting when asked about it in late January 1997. "All the good people are so busy with F1," he said haltingly with a pained look. "It's such a big effort that we don't have the capacity in terms of people. Nor do we have the financial capacity for another activity."

An insider close to upper management said that wasn't the cause because, "The funds for the project came from Brunei. The Sultan paid for the car's development work, and three F50 GTs were supposed to go to him. What caused the program's cancellation was Montezemolo said no. There were fuel regulations then, and he felt Ferrari would get beat because the F50 GT had a V-12 and would lose too much time in the pits refueling."

Like Piero Ferrari, Franco Meiners said the real culprit was finances, but not in the manner everyone imagined. "I was really annoyed by the program being shut down because everything to prepare the car had been done," he recalled. "We were ready to race, and I wanted to know the real reason. There were so many reasons swirling about [so] I did some digging and was told that, around the time of the cancellation, the Fiat Group had asked for financial intervention from the Italian government because they needed money. I think it was something like 400 billion lire, and the government replied, 'Okay, we pay the money but it is only for the company and the workers and is *not* to be used to go racing.'

"This condition was attached because it was still 1996, and in the beginning of 1997 there were going to be elections. The government was thus scared to death to see people in the middle of the road in December before Christmas, protesting what had happened.

"When they said 'no racing,' this meant the Fiat Group had to close down all the competition departments, of which one was Ferrari's F50 GT. Obviously, Formula 1 is constitutional so you can't shut that, plus it also had separate sponsors and so on. But everything else had to close."

Nearly two decades later, one can sense Meiners' lingering frustration. "The F50 GT was an absolutely fantastic car, so well balanced and sorted," he said. Then he started thinking about chassis 001 and laughed as he asked, "Where is the Dallara car? It is great, *really* something!"

That first car was sold off to prominent Ferrari client and collector Art Zafiropolo under the provision it could never be campaigned and sent to him in California in April 1997. As of this writing he still owned the car.

In 1996 and 1997 Dallara also made chassis 002 and 003. Like 001, they were sold to collectors under the provision they never be raced.

To put an exclamation point on the model's potential—and after nearly two decades of considerable advancements in electronics and active aerodynamics—the F50 GT still remains the fastest non-F1 car to lap Fiorano.

⌃ No sooner had the F50 GT's preliminary testing been completed than the program was canceled. As seen in the text, numerous theories abound on what happened. This is F50 GT 003 under construction in 1997 after the program's cancellation.

⌄ One reason the F50 GT was so blisteringly fast was its potent powerplant. Though Ferrari lists power output at 750-plus, those with knowledge of the first car say the number may be closer to 1,000.

SECTION III:
THE DIGITAL AGE
(1998–PRESENT)

"I do not know what will become of my works after I die, but to my successor I bequeath a very simple inheritance: to keep alive that constant striving after progress that was pursued in the past."

—Enzo Ferrari

According to a number of people interviewed for this book, Ferrari underwent a tremendous revolution in how it designed and developed its road cars in the mid-1990s. One of the main reasons for the change was the 348; when the model was launched, it had extremely twitchy handling and inferior build quality and materials.

"WHEN WE WERE DOING THE F50," Lorenzo Ramaciotti recalled, "Ferrari was undergoing a turbulent time because it was moving from the Enzo Ferrari period to the Luca di Montezemolo period, and [Amedeo] Felisa was coming."

CORNERSTONE TO A REVOLUTION

The name "Felisa" may not have much resonance outside Ferrari circles, but this incredibly talented engineer has been the quiet cornerstone of a design and development revolution that swept through Maranello.

Focused, jovial, and possessor of a wide-ranging skill set, Amedeo Felisa was born in Milan in 1946 and has been a gearhead for so long that it's an indelible part of who he is. When asked in 2006 what hobbies he enjoyed, he replied, "On the weekends I try to relax, and I read a lot. Also travel. These are my passions. And cars, of course."

Felisa graduated from Milan's Politechnico College in 1972 with a degree in mechanical engineering and was hired by Alfa Romeo to work in their experimental department. "I stayed until 1984," he told journalist Matt Stone in the Ferrari specialist magazine *FORZA*. "In those times, it was difficult to understand what was to be the future of the company. I decided to move on and went to a big Italian company that made agricultural machines. …When Fiat bought Alfa in 1986, they asked me to come back. Then they asked me to [go] to Ferrari, essentially an internal transfer, and I said, 'Why not?'"

That was 1990 and with his arrival at Ferrari came an entirely different way of doing things. "His approach was not to be the guy who was designing the car but the guy that built the team of talented engineers working for him," Ramaciotti explained, summarizing what a number of others said. "He gave them the responsibility and the chance to express themselves."

This new methodology has seen Ferrari continually update and revamp its model range at a rate unimaginable in the Enzo years. "If you are here as an engineer," Felisa pointed out to Stone, "you have so many technical enhancements to do every month—a lot of projects, all the time. I made more cars in the first four years after I came here than

≫ Though the 348 became a very good car by the end of its life cycle, the 355 was an altogether different machine. It was the model that served notice that Ferrari had once again returned to that special universe it seemed to occupy.

≫ So who is responsible for the radical metamorphosis that Ferrari's road cars have undergone since the early 1990s? During research, four people independent of each other pointed to this man: Amedeo Felisa, an engineer whose name is largely unknown outside Ferrari's inner circles. *John Lamm*

I had in fifteen years with Alfa Romeo. This is the best. We move fast and have probably averaged at least one new car development project per year."

He's a believer in strategic outsourcing. As he explained in 2003, "Sixty percent of the effort involved in making a car involves suppliers, so I spend a lot of time with them. We use the suppliers where they have development skills we don't have, or where they're just better than us."

THE SPEARHEAD

That creative energy needed a target, and the man who has chosen it since the early 1990s is Luca Cordero di Montezemolo. A rock-star personality in Italy, Formula 1, and auto industry circles, in many ways he's been the second coming of Enzo Ferrari, as brash as that sounds. "Enzo might have created the Ferrari legend," *Motor Trend*'s Angus MacKenzie wrote in 2011, "but it's Luca, born into an aristocratic family the year the company was founded, who's made the legend real."

While that's stretching things a bit, there's no disputing Montezemolo made Ferrari what it is today—one of the most recognized names worldwide and likely *the* automotive benchmark for performance and engineering.

A fiercely competitive individual ("He's got a passion that his competitors can feel, and sometimes it can be uncomfortable to be around," McLaren's Martin Whitmarsh told the *Wall Street Journal*), Montezemolo was born on August 31, 1947, into a wealthy, powerful family from northern Italy's Piedmont region. Throughout his life he'd have Fiat patriarch Gianni Agnelli as a mentor ("He was the most important person in my life outside my family," he would recall) and was friends with other Agnellis.

It's hard to imagine a better name to drop in Italy's motoring world than "Agnelli," especially when having the chutzpa to back it up. His first stint at Ferrari came in the 1970s, and how he got the job is unique to say the least. He'd raced on the Lancia Rally team, so while studying law at Columbia University in December 1972 he was invited to be a guest on *Chiamate Roma 3131*, a popular Italian radio talk show.

A caller began heckling him that his rich-kid background was the reason he was driving, so Montezemolo fired back forcefully. The way he responded piqued the interest of one listener in particular—Enzo Ferrari. As Montezemolo recounted years later in the *WSJ*, "Ferrari called [the radio show] and asked, 'Who is that tough young kid with the balls to answer that idiot?'"

Enzo then approached Montezemolo, telling him he needed someone like him to reorganize the racing department. Montezemolo's response was simple: "You don't have to ask me twice." His family vehemently objected, but Luca didn't listen. At the ripe age of twenty-six, he was running the world's most famous racing team.

His first year, 1973, saw the Scuderia win just two races. The following year Niki Lauda came on board, and the transformation hit high gear. Lauda won the drivers' championship in 1975 and '1977, and Ferrari swept the constructors' titles from 1975 to 1977. "[Montezemolo] took the job that everybody thought was impossible," F1 veteran John Hogan observed in the *WSJ* "His fingerprints were everywhere on the team; you could see he was going to be a power broker."

Which is exactly what he became. Montezemolo was appointed the Fiat Group's director of external relations in 1977 and four years later became CEO of the Fiat Group's publishing division. For the second half of the 1980s he headed *Itala '90*, which organized Italy's successful hosting of soccer's World Cup finals. He also had a stint as head of aperitif maker Cinzano and spearheaded Italy's foray into the America's Cup.

In November 1991 he returned to Ferrari. "Those were difficult years," Montezemolo told Italian journalist Luca Cifferi, "with the market in crisis, the project drawers empty, and a certain arrogance that it was enough to be called Ferrari to be successful."

As if to emphasize the point, he often tells the tale of purchasing a 348 in 1990 and being thoroughly unimpressed. Not long after his arrival in Maranello, as the engineers raved about their then-current lineup, he cut the meeting short, bluntly telling them "the 348 is not a Ferrari for me."

Therein lay one of his greatest strengths: He puts himself in the mindset of a typical Ferrari buyer because he himself *is* one.

Another has been his ability to transform the firm: For the past decade-plus Ferrari has been anything but a provincial exotic car builder ruled by an autocratic leader. Instead, it's become a progressive organization with sprawling grounds, spacious streets, several *Architectural Digest*-quality buildings, and recipient of "The Best Place to Work in Europe" in 2007. Motezemolo told *The Financial Times*, "I strongly believe if you want the best product you need the best organization and people, and you must put people in the best condition to work."

"FROM THE OLD TO THE NEW"

Once Montezemolo was situated in Maranello, Felisa and his fellow engineers were turned loose. "The 348 was a car that was almost impossible to drive because it was very sensitive to the stance, to the ground position, so every car was behaving differently," Ramaciotti recalled. "These guys, without changing the mechanical layout, just working on the characteristic angles and the bushings, transformed [it] into the 355, which is a fantastic car—very entertaining with no problems, and very quick. They did the same with the Testarossa, changing it into the 512.

"What they did was alter the approach from one where [it was] 'we make the layout and the suspension and we test and see how it works' to one where they started making simulations, making use of forecasting tools, CAD, and so on. It was really a move from the old to the new. . . . "

Felisa explained it this way to *CAR*'s Paul Horrell in 1996: "The work we've done on the F355, 512M and F50 is to make the car 'sincere,' so that you feel when [it] is moved to spin. This is most important. The car must tell you, 'I am going,' and you need the right moment of inertia so you can control it. If it's too low you'll have problems unless you're a very skilled driver. To make it sincere, we manage the ratio of wheelbase to track, the geometry, the tires."

Another philosophical change was, "Ferrari has to be thought of as a special car, but still a car," Felisa told Matt Stone. "A big reason for our success . . . is that you can really use a Ferrari now, every day if you want, not like in the past."

« In 1998, Ferrari revolutionized F1 with its hydraulically actuated paddle-shift transmission. The first road car to get the technology was likely the Ferrari FX, made in secret in the mid-1990s for members of the Brunei Royal family. The first production car to get the technology was the 1997 355F1.

The 355's debut in 1994 highlighted these advances. "I'm very pleased to say that Ferrari has succeeded with . . . the 355," Gavin Green's editorial stated in *CAR*'s July 1994 issue. "Its predecessor, the 348, was not the most competent car of its type. The NSX and new 911 were more faithful, faster, more maneuverable. The new F355 not only beats its rivals in brio and beauty, but it also beats them at the important business of serving up a fast yet sure-footed drive. . . ."

Paul Horrell's sum up of the issue's cover story was this: "After [spending] two days with [a 355], Ferrari has moved the whole game forward. . . ."

That model was simply Maranello's opening act. An industry-wide horizon that started with ABS in the 1980s was opening up, one where advancements in electronics made mechanical components work more efficiently. Ferrari traveled down that road with a vengeance, developing a game-changing technology based on the system that revolutionized F1 several years earlier: the electro-hydraulic paddle-shift transmission.

Introduced in the 1998 355F1, "I can already hear you thinking: Who the hell wants an automatic Ferrari?" was how *Motor Trend*'s May 1998 comparison test summarized their experience with the car. "Well . . . it's more like having an invisible Michael Schumacher onboard to up- or downshift at the precise moment, blipping the throttle to match revs, and squeezing every bit of performance from the delicious . . . V-8."

MT found the 0–60 and quarter-mile times were two-tenths of a second quicker than a manual transmission car, and pedalless shifting became immensely popular with Maranello's customer base. Indeed, in not much more than a decade Ferrari's famed gated shifter was a thing of the past.

"It seems everyone has an uncle who was the start." That's how former Pininfarina and current Fiat design head Lorenzo Ramaciotii described what sparked his unbridled passion for cars. He was the only child of a Modenese banker father and housewife mother, born on January 6, 1948.

About that uncle: "He was a real car enthusiast who had sports cars like the Lanica Aurelia B20. He also had an Alfa 1.9 and was changing the car quite often. In so many families there is an uncle who had some special capacity that was very fascinating for a boy—perhaps he took them fishing or hunting—and I had this kind of uncle."

As if to add mystique to the impressionable youngster, the uncle passed away at the age forty-two, "so he was vanishing, kind of like a James Dean. . . ."

In the short time Lorenzo knew him, the uncle obviously made a strong impression: "I have an aunt who is ninety-eight, and she reminded me recently that when I was a boy of three or four, I was recognizing the sounds of cars passing by. This is an art that was much easier back then because there were only a few cars around, plus they were making noises quite different from one another."

Ramaciotti went on to have what he called "a standard scholastic career" and graduated with a degree in mechanical engineering from Turin Polytechnic. For work, "I sent my curriculum to many of the companies around, Pirelli, Fiat, and so on, but in the end Pininfarina hired me. I wanted to be a racecar engineer, as my idol was Colin Chapman. My dream was to be Italy's Colin Chapman, but I had the chance with Pininfarina, and I think it's important to grab a chance when it comes. So I [did]."

Like many in the auto industry, he remembered the date of his first workday: January 2, 1973. "When I had my interview," he said, "I told them I was only interested in working in the research department, and that I wanted to design cars. Maybe they hired me because of that, for I didn't seem interested in just any kind of

Immersed in the nerve center of leading automotive design, Ramaciotti discovered his true skill was not in putting the pen to paper but assembling a team and successfully guiding it. "I'm not a soloist," he explained, "and the work we do is very complex. When I started it was already like that. Maybe it was possible [to be a soloist] in the Michelotti or Scaglione times, I don't know. More and more [design] is not like . . . work done by a single hand; it's the play of a team, where you have to have the right talents at the right place and have them working together, like a football or basketball player. You need a coach who puts him in the right place on the field.

"There are so many positions in our [field of] creativity, and a fantastic designer can have success if he has someone giving him the right direction and the chance to play his role."

That philosophy, and a passionate work ethic, served him well. He soaked up lessons imparted by Fioravanti ("I owe him a lot, because he really gave me the right direction"), and was made a deputy manager in 1982. In 1988 he succeeded Fioravanti to become the design studio's director, and oversaw more than 100 designs over the next seventeen years.

Notable production Ferraris under his watch include the 456 (1992), 355 (1994), 550 Maranello (1996), 360 Modena (1999), F50, and Enzo. Other highlights include the Ferrari Mythos (1989) and Rossa (2000) prototypes, Peugeot 407 coupe (1995), the Maserati Quattroporte (2003), and Birdcage.

He retired in June 2005, only to return to work in 2007, when Fiat asked him to oversee six of its divisions. On why he went back to being a design director: "They made me an offer I couldn't refuse," he said with a grin.

Now six years into his "new" job, it's obvious he still relishes the work. "I wouldn't be here if I didn't like it," he quickly pointed out. "Every car has it's own basket of constraints. That means every basket is different, so it is interesting because you're not facing the same thing."

« ≋ Over the years Ramaciotti and his design teams have shown great diversity—from the classically beautiful Ferrari 456 GT that debuted in 1992 (shown is the later M456 GT variant) to the 2000 avant-garde one-off Ferrari Rossa concept (left).

≋ Ramaciotti is totally into—and a great student of—car design and especially automotive history. He's long served as the chief judge at Villa d'Este, one of the world's great concours d'elegance. At a 2006 event, he's seen giving some of his typical insight from the judge's podium.

About the incredible transformation Ferrari's road cars underwent in the mid-1990s, "What they did was change the approach," Lorenzo Ramaciotti noted. "They started making simulations, making use of . . . CAD, and so on." Two of the first Ferraris to benefit were the F512M (foreground) and 550 Maranello.

PERFECT TIMING

The F1 system's "go faster, easier" mentality was right in tune with the late 1990s resurgent interest in performance cars, wealth, success, and the world's burgeoning fascination in new, cutting edge technologies.

CAR did two special issues in 1997 with succinct cover titles that highlighted the first two trends: "SPEED!" (April) and "That's RICH" (November). Over in California, Silicon Valley was booming as commercially applied science attempted to make life easier and more convenient. "Dot-com" entered the global lexicon, mobile phone sales were taking off in new markets, and "connectivity" was becoming prevalent in societal consciousness everywhere. America's technology-laden NASDAQ Composite Index shot up like an arrow, hitting an all-time intraday high of 5,132.52 on March 10, 2000, barely five years after the index crossed 1,000 for the first time ever.

That and more made the era the ideal time to create what, up until recently, was one of the most advanced performance cars ever. That machine was the Enzo, and at its core were three tenets.

"First," Montezemolo told Steve Cropley in *Autocar*'s *Ferrari Enzo* booklet, "I wanted it to be something really impressive. Second, I wanted it to be technically the best we could do. Third, I wanted it to be a celebration of our racing success, but not so close to racing in its design that the relationship hurt its performance as a road car."

In short, the Enzo wasn't simply to be Ferrari's next "ultimate"; it was to make a *statement* about the company's capabilities. "I wanted to go a little too far in every element," Montezemolo told Cropley. "I said to Felisa we have to put on this car, with our key partners, the best we can do, all our overall know-how from racing and road cars. If we can't find it inside we'll call around."

"I said to Sergio [Pininfarina], 'Go too far, it's easy to come back . . . I want an extreme car.'"

— Luca Cordero di Montezemolo

Inside Maranello the project was simply known as the FX, and much like Piero Ferrari did in the early stages of the F50's development, Amedeo Felisa organized lunches between key personnel in the F1 and road car departments. As the engineer noted in 2002 to *CAR*'s Mark Walton, "Every time we finish a car like [the Enzo] we wonder how to do a new one. One year after we finished the F50, we could see how we could build a better one."

Enumerating further, "We knew the weak points, the mistakes of the F50," Felisa told *Autocar*'s Peter Robinson. "The engine was too small, the noise too high. And technically, for a car of this performance, we knew it was better for the FX to be a coupe."

In late spring 1998 those lunchtime conversations progressed far enough that the group had determined the FX's key parameters, so Montezemolo called Sergio Pininfarina to prime the design pump. As Ferrari's man recalled for Robinson, "I said to Sergio, 'go too far, it's easy to come back. . . . I want an extreme car."

Pininfarina welcomed the assignment. "He always had a special attitude toward Ferrari cars," Ramaciotti recalled. "He didn't come all the time [to meetings] for the development of other companies' models [but] with Ferrari, he was always present at the presentations. If he had to miss one, he made sure everybody was saying that he was very sorry not to be there. He wanted to be informed about everything, so it was a personal sweet spot.

"Of course, Montezemolo trusted him because he was a part of the Ferrari history, a very nice guy with a lot of taste, and a huge reputation worldwide. With 'Ferrari by Pininfarina,' it was already a kind of signature. The two guys had a very strong relationship [even] though there was an age gap of more than twenty years. Montezemolo was asking Sergio, 'What do you think?'"

In early June 1998 a briefing was held at Pininfarina's Studi e Ricerche design facility in Cambiano, where Ferrari's Giuseppe Petrotta laid out the parameters. The gifted engineer's career began in the late 1970s at Osella, where his work on their prototypes and F1 cars saw him promoted to chief designer for the F1 team. From there it was to Alfa Corse before he came to Ferrari in 1991. His first two years were in the F1 team, then he began engineering the road cars; the 360 Modena's aluminum space frame was an example of his work.

At that first meeting, Ramaciotti recalled, "The concept was 'we must break the mold of the F50' and do something worth having. All the customers of the F50 should say, 'This is old stuff, and I have to have the new one!'

"We tried to [do this by making] first, not a stylish car aimed to please, but one that looked very purposeful. It had to be strongly recognizable without compromise. . . . We wanted it very crude."

An internal competition was set up in Pininfarina, the designers competing for what remains arguably the most coveted assignment in the automotive world—designing Ferrari's next hypercar. Sketches were made that portrayed numerous themes and details and were continuously presented to Montezemolo and Petrotta in a series of review meetings.

Then came the presentation that would transform the FX "idea" into what became the actual Enzo. "We went to Maranello with a scale model we were developing and put it in their showroom," Ramaciotti explained. "We had at least three, and all were placed on pedestals. We made the presentation to Montezemolo . . . and he was, 'Yes, but I want something more Formula 1, more extreme.'

"There was a huge poster on the wall, larger than life of Schumacher with a Formula 1 car. At the time Formula 1 cars were not as swoopy as today; they were much more dramatic. We said, 'If you want Formula 1 on the road, we must do *that* car. We went back [to Turin] and said . . . let's *really* do the shape of a Formula 1 for the street. That's why [the Enzo] is—instead of being let's say sensual and sculptural—much more dramatic. That was the technology of Formula 1 at the time."

MIND MELD

That the Enzo's radical shape showed such "certainty" leads one to conclude the creative process was easier than the F50. Ramaciotti makes it clear this wasn't the case.

"Montezemolo put the bar very high," he explained, a subtle grin playing across his face. "He always wants the new Ferrari to bring something more than the old one. During the project, many times he challenged the team, when you are well established in [what you had done]. He sometimes comes in and says, "I don't like it! *Non piace piu*! Now we want to do something stronger.'. . .

"I think this is one of Montezemolo's strengths: He tries to bring out the best of the people who are working with him in terms of participation, by having a strong conviction in what they are doing. If you have to convince someone else, you yourself have to be really convinced in what you are doing."

The Enzo's design would be set in stone in January 2000 but not before Montezemolo "did his stuff one more time." As Ferrari's leader confronted the design team over the scale model that would become the production car's form, Ramaciotti recalled, "Sergio Pininfarina told him, 'Listen, Luca, I really believe in this car. It has a lot of potential, a lot of strength. If you won't do it for production, I'll be using it as a concept car for the next show.'

"Montezemolo found Pininfarina so strong on this . . . that he said, 'You have convinced me. We go on with this.'" (In fact, Pininfarina was *so* certain that at the Enzo's formal unveiling at 2002's Paris Auto Show, he called its design "the third quantum leap" in Pininfarina Ferraris, after the 250 SWB and Dino.)

» Montezemolo and Sergio Pininfarina had great respect for each other. When Ferrari's CEO challenged Sergio's design team one last time on what would become the Enzo's shape, "I really believe in this car," Pininfarina replied. "If you won't do it for production, I'll be using it as a concept car for the next show." *Pininfarina archives*

One of the toughest elements to resolve was the doors. "The problem with these [type of] cars is the wheels are huge, and the wheel arches are really big so you need to have the two people sitting side by side," Ramaciotti explained. "In this case . . . we put them really close [which made for] a very small cockpit. . . . To get into the car you have to come over a very wide sill, where you slide between the roof and the sill. We needed better access, and since the tub is much narrower than the body, I felt [we should] take the sill away."

Ramaciotti would make a superb automotive historian or pundit if he weren't so content as a talented design director. He felt such cars needed to be unconventional in unexpected ways, and recommended the FX have "a much more visually entertaining door opening [because] there is a huge part of the body that is attached to the sill. [This way], instead of slipping over the sill, you stand close to the seat and enter much more easily."

He conveyed his solution by noting, "The whole door should open to the ground, as the Citroen DS was doing that. But the [FX] was so low that if you do that while parking near a curb, you couldn't open the door. So we said it must open upwards [to] break away the whole sill and part of the roof. We saw pictures of the old Ferrari [512] racing cars that did this, [and] used that butterfly [configuration]."

Once everyone was in agreement, Ferrari became nervous about the doors' potential complexity. "You need a counterbalance to hold the stuff up when it is open," Ramaciotti pointed out. "They said, 'We'll do it for it's a breakthrough, gives a lot of people interest, and is very comfortable, though we'll face a lot of complications.' But at the first trial the door was working and it never gave complications! It was perfectly engineered by the guys who were engineering the car."

⌃ Enzo chassis 129581 is seen on the Pininfarina stand during the model's official debut at 2002's Paris Auto Show. The Pininfarina designer responsible for the car's aggressive shape was Ken Okuyama.

⌃ On Ferrari's stand next to an F1 car is Enzo
chassis 128778, while on Pininfarina's stand is
Enzo chassis 129581.

FERRARI ENZO

Year(s) made: 2002–2003
Total number produced: 399+1
Hypercars: all
Drivetrain: 5,998cc DOHC V-12, 660 hp @ 7,800 rpm; six-speed
manual paddle shift
Weight: 3,230 lb (U.S. spec)
Price when new: $643,330

PERFORMANCE:

0–60 mph: 3.3 seconds
0–100 mph: 6.6 sec
1/4 mile: 11.1 seconds @ 133 mph
Top speed: 218 mph
Road tested in: *Road & Track*, July 2003
Main competitors: Porsche Carrera GT, Mercedes, SLR, Pagani Zonda
C12S, Saleen S7

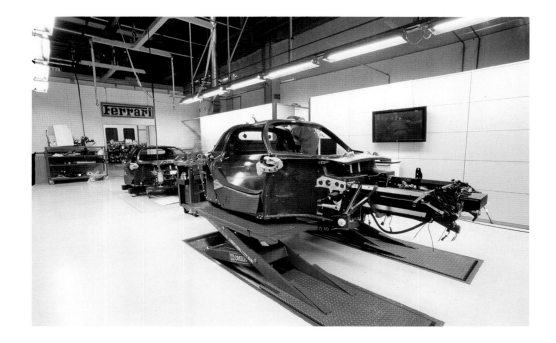

« Not too long after the Paris Auto Show, Enzo production began in Maranello on a line dedicated to the model. Here Enzo carbon-fiber tubs occupy the line's first two stations on February 25, 2004.

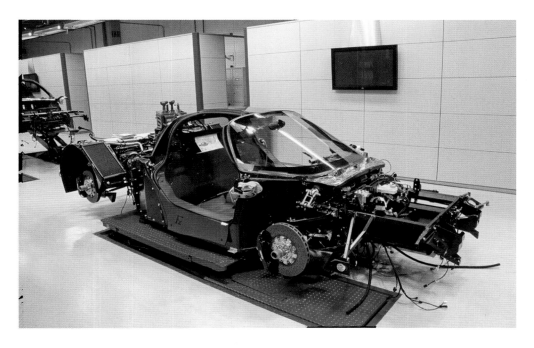

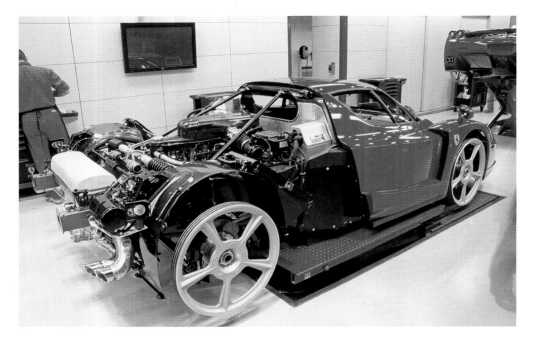

« ⌃ On December 3, 2002, chassis 131024 is in the middle of the eight assembly stations on the line. A peek inside the chassis tub 131024 highlights the carbon-fiber weave.

« On December 3, 2002, Enzo chassis 131023 awaits its rear quarter panels and deckled.

≈ Enzo chassis 130918 has most of the bodywork in place.

UNDER THE CARBON SKIN

In Maranello, Felisa and Petrotta's tightly knit development and engineering group started with ten people and grew to fifty. "We had a list of internal targets, including performance," Petrotta told Robinson. "But we also knew it was a technology laboratory for future Ferrari road cars."

Though no technology was off limits, a large hurdle was constantly evolving government regulations. From numerous computer simulations, Ferrari devised a rigid carbon-fiber and Kevlar monocoque structure. The roof was bonded to the tub, and the engine sat in a special cast-alloy subframe cradle; the idea was the cradle would eliminate the noise and vibration issues that so concerned everyone at the F50's launch. The two fuel tanks were made of alloy and housed in a cabin between the engine firewall and driver's compartment.

The team never considered anything but a naturally aspirated V-12 and targeted an output of 630 horsepower. The engine was codenamed the F140 and, after examining vee angles from 60 to 75 degrees, they settled on 65 degrees, as seen in the F50. The engine was placed longitudinally in the cradle and featured alloy block and heads, DOHC, four valves per cylinder, titanium connecting rods, and a trick lightweight crankshaft. The cylinder bore was 92mm, the stroke was 75.2mm, and a 1995 F1 engine was the basis for the variable length induction system. The throttle was drive-by-wire, and the continuously variable valve timing was a first in a street Ferrari. Two Bosch Motronic ME7 ECU's controlled the six anti-knock sensors.

In final trim, the 5,998cc V-12 produced 660 horsepower at 7,800 rpm, and 485 lb-ft at 5,500 rpm. Redline was 8,000 rpm. The six-speed sequential manual gearbox was mounted behind the engine and used a twin plate clutch. Paddles behind the steering wheel took care of the shifting, and an F1-inspired launch control program was included to ensure perfect high-performance starts from a dead stop.

The car's three-spoke steering wheel brought additional F1 technology and ergonomics to road cars. Packed in the center hub were six buttons; these allowed the driver to control functions such as altering the gear changes, ride height, and traction control. The turn signal controls were in the wheel's spokes, and a series of small LED lights on the upper portion of the steering wheel rim signaled 5,500 rpm to 8,000 rpm in 500-rpm increments.

The suspension front and rear had double wishbones, pushrod links, and coil springs over adaptive shocks. This last item reacted to four accelerometer sensors on the car's body, a speed sensor, two vertical wheels sensors, and the brake switch setting.

The 380mm ventilated disc brakes also broke new ground. Whereas carbon fiber was avoided on the F50's brakes because of cold temperature concerns, the Enzo used Brembo's revolutionary Carbon Ceramic Material. "[The brakes] are fantastic," test driver Dario Benuzzi told Steve Cropley as the technology underwent final testing. "If you consider the Enzo weights 800 kilograms [1,616 pounds] more than the F1 car, these brakes are better than the F1. There's no vibration on the track, no variation in their performance hot or cold. They are the best."

But the real parlor trick of the Enzo, one that showed the digital age had truly come to cars, was the way all the electronic systems "talked" with one another, rather than working separately. As Ferrari's aerospace engineer Stefano Carmassi told Peter Robinson, "From the beginning the accent was on thinking of the car as a whole, not a composition of different subsystems. We wanted the best possible behavior in all conditions."

The engine, transmission, anti-lock brakes, ASR traction control, suspension, and active aerodynamics were completely integrated, seamlessly working in concert with each other to maximize performance as never before. For the active aerodynamics, the team determined that enclosing the underbody was essential. "I wanted to have this Formula 1 nose with this tunnel in the middle to feed the underbody," Ramaciotti said. "We [also decided] we didn't want to have any wings in the back, [for the shape] to be as smooth and integrated as much as it could."

They succeeded. A major benefit was the front and rear aero balance could change, as the rear spoiler continuously acted in unison with flaps ahead of the front wheels. Other advances included splitting the underbody airflow with a central diffuser to generate more downforce since the air was flowing freely from the front to the rear.

"It's easy to build a car capable of 300 kilometers per hour [186 miles per hour]," Carmassi told Robinson. "We wanted a car whose character allows you to drive it at a useable 300 kilometers per hour."

"I WANT TO DRIVE IT NAKED"

As prototypes underwent development, Michael Schumacher brought his considerable skills to the process. "I drove it at Fiorano to check the drivability, the flexibility, of the engine and gearbox," the F1 ace told *Autocar*'s Peter Robinson. "I try to feel what the car is doing on the road to my standards. This does not mean in racing conditions [as] this is only a small part of our interest."

When Schumacher performed on-road testing, the most nervous people in Maranello were found in the racing department. "[They were] worried about Michael's [driver's] license," Felisa told Robinson. "They didn't want any trouble with the police. He [once] came out of [a corner] like this [Felisa demonstrated by crossing his arms in classic oversteer style], where a police car had stopped another driver. We didn't think they could catch us so we kept going."

In summer 2001 spy photos were seen in magazines, and in April 2002 the car had a surprise unveiling at the Ferrari/Maserati exhibit at Tokyo's contemporary art museum. Maranello released a single front three-quarter photo of the car, some publications reporting the upcoming hypercar would be called the "F60."

⌄ Overseeing Enzo (and F50) production was the talented and gregarious Maurizio Moncalesi.

On the production line on December 2, 2002, the closest Enzo is chassis 130728 that was destined for the United States. On the lift is chassis 130918. The third car is chassis 131022, while the last nose visible is chassis 131023.

The official public debut took place at the 2002 Paris Auto Show, but several weeks before a small group of journalists were invited to Maranello to drive the Enzo for several laps around Fiorano. *CAR*'s Mark Walton was so enthralled by the experience that he started his road test relating how, right after he settled into the driver's seat, he turned to Dario Benuzzi and baffled him by saying, "I want to drive it naked . . . butt naked, except for a pair of leather driving gloves."

"The Enzo is like a monster [Lotus] Elise, so delicate at the controls, yet punishingly violent out of corners," Walton marveled. "Onto the straight and open it up through three gear changes, red lights blinking on the steering wheel, it starts to feel like a 747 at take-off speed at the end of a runway."

Two of the car's greatest attributes were the brakes' unparalleled stopping power, and the car's unobtrusive NASA-like brainpower. About the latter, Walton noted, "Even coming out of the tighter turns with caution, the traction control light is blinking, hinting at the gargantuan forces at work here. Yet when you drive it doesn't feel like the electronics are interfering. It just makes you feel powerful, deep in your inner psychology; you wield vast power at the wheel of this car, and you start to feel like it's you, reaching out and grabbing the next corner, wrenching toward you. . . .

"Where the McLaren F1 made drivers feel fear and awe, the Enzo Ferrari makes you feel fearless and awesome."

THE BURNING QUESTION IS ANSWERED

Walton's praises were summed up in the issue's cover headline "NEW! The World's Greatest Ferrari," while the subhead tackled the 800-pound gorilla in the room: "But Can It Beat McLaren's F1?"

That's what everyone wanted to know, for several years earlier the McLaren had widely been hailed as the greatest, fastest performance car ever. Indeed, *Autocar*'s memorable May 11, 1994, road test had a caveat much like Walton's "fear and awe" observation: "This is a car which unless driven with a cool head, could land in you in greater trouble than you could imagine. As we said, you need to pick your moments with the F1."

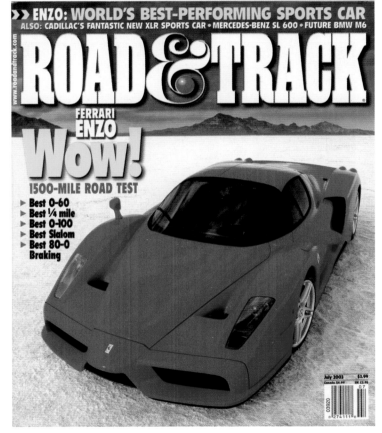

That's where the Enzo represented a true breakthrough, for it completely altered the driving equation. During the analog years, man and machine were true partners, one basically as important as the other. In the digital age, the Enzo's incredible software and systems' seamless interaction largely masked and thus offset a driver's skillset deficiencies.

Road & Track's July 2003 cover answered the Enzo/F1 debate by declaring the Ferrari "The World's Best Performing Sports Car." When compared to a Federalized F1, the Enzo's 0–60 time of 3.3 seconds was one-tenth of a second quicker. The gap was larger to 100 miles per hour (6.6 vs. 7.7 seconds), and in the quarter mile (11.1 at 133 miles per hour vs. 11.6 at 125 miles per hour). Braking distance favored the Ferrari again: 109 feet from 60 miles per hour (127 feet, McLaren) and 188 from 80 miles per hour (215 feet). Lateral acceleration and the slalom was much the same: 1.01g on the 200-foot skidpad (0.86g, McLaren), and 73.0 miles per hour on the 700-foot slalom (64.5 miles per hour).

Only in top speed was the McLaren superior (231 miles per hour vs. 218). But with speeds in excess of 200 miles per hour becoming more common in the digital age, that yardstick (especially after the arrival of Bugatti's 1,000-horsepower, 254-mile-per-hour Veyron) was no longer the real determinant for anointing something "the greatest."

⌃ The inspiration for the Enzo's butterfly doors, where it hinges forward, was 1970's 512 S. Cutting away the roof as the door swings open allows one to step over the tub's wide sill and enter more easily. Plus, such road-going ultimates should have what Ramaciotti called "a more visually entertaining door opening."

"The myth that a super sports car with exceptional performance cannot be civilized at the same time has been broken," *R&T*'s Patrick Hong summed up after the editorial team completed its 1,500-mile multiday road test through Utah. "The Ferrari Enzo is both an ultra-high performance car and a capable grand tourer. . . . Maranello has entered the Enzo into another supercar realm where the rest of the world has to catch up."

Autocar's Steve Cropley reached the same conclusion the first time he drove an Enzo: "We have all known that somebody, someday, would build a better car than Britain's McLaren F1. Ferrari's Enzo is now that car. It's no faster in a straight line, but it's handling, road-holding, steering, brakes, and even ride comfort, are all comprehensively better. Ferrari has built the new supercar benchmark and given it the greatest name of all."

WHEN AN ENZO IS NOT AN ENZO

A total of 399 Enzos were made in 2002 and 2003. A number of months later Ferrari constructed one final car for Pope John Paul II that was auctioned off, the proceeds going to tsunami relief.

At the time that happened, Ferrari owned 100 percent of its once mortal enemy Maserati, so a decision was made that Maserati would get its own mid-engine missile. Given the two companies' lengthy histories, it's most appropriate that that chapter of the Enzo story largely took place on the racetrack.

Maserati's MC12 was originally called the MCC and was built on the Enzo production line after a temporary wall-like barrier was erected to separate it from the normal production lines. The car's wheelbase was nearly 6 inches longer than the Enzo's, overall length was 11-plus inches greater, and the width and roofline were 2-plus inches more.

OPPOSITE PAGE: That Formula 1 played a huge role in the Enzo's shape is easily discerned from a higher angle. The raised section in the center of the front clip clearly mimics the architecture of a late-1990s F1 car.

With most Enzos red, perhaps the most stunning of all is chassis 135872. The smoky metallic silver
(Grigio Titanium) and red interior is a unique combination on the model, the exterior color really
showing off the model's architectural line.

≈ A popular interior was this red and black combination, as seen on chassis 135872. The circle near the grab handle is the window crank.

Enzo 135439 on the move. *Motor Trend*'s test Enzo recorded 0–60 in 3.1 seconds and the quarter mile in 10.8 seconds at 135 miles per hour—still impressive numbers more than a decade after the model's debut.

The chassis remained carbon fiber, and the suspension was similar to the Enzo's with pushrod actuated coil-over shocks. The engine was modified for race duty, and featured gear-driven cams, an upgraded dry sump lubrication system, and different cams and pistons. Everything—including the ECU—was made to optimize breathing through a mandatory 33mm intake restrictor.

« The MC12's real raison d'être was Maserati's return to international competition. That it did with a bang, with the MC12 GT1 (seen here in Bahrain in November 2005) winning numerous FIA GT championships from 2005 to 2010. *LAT*

Ferrari design head Frank Stephenson created the Maserati's stretched fluid shape, and overseeing the car's production was the gregarious and talented leader of the F50 and Enzo line, Maurizio Moncalesi. The MC12's competition debut occurred at Imola on September 4, 2004, where the two cars finished second and third OA. They then won the next two races.

For 2005, the car was developed further and known as the MC12 GT1. It featured a 745-horsepower engine and, according to *CAR*, could lap Fiorano in 1:13. The MC12 would dominate the FIA GT series for the next five years, winning fourteen championships (two constructors', six team, and six drivers').

MC12/MC12 GT1, MC12 CORSA

Year(s) made: 2004-2005 (MC12 & GT1), 2006 (Corsa)
Total number produced: 50–55 (MC12), 15 (Corsa)
Hypercars: 25
Drivetrain: 5,998cc DOHC V-12, 624 hp @ 7,500 rpm; six-speed manual paddle shift (MC12)
Weight: 3,150 lb
Price when new: $792,000

PERFORMANCE:

0–60 mph: 3.8 seconds
0–100 mph: 8.0 seconds
1/4 mile: 11.8 seconds @ 123.9 mph
Top speed: 205 mph
Road tested by: *Motor Trend*, June 2005
Main competitors: Pagani Zonda F, Saleen S7 Twin Turbo,

OPPOSITE: The engine compartment of Enzo chassis 135439. The 5998cc V12 produced 660 horsepower, an astronomical amount at the time of introduction. Suspension had double wishbones, pushrod links, and adaptive shocks that reacted to accelerometer sensors.

Depending on the source quoted, fifty to fifty-five MC12s were built in 2004 and 2005; this included 25 targa-topped road cars. *Motor Trend* tested a 624-horsepower road-going version at Alfa's Balocco test track in its June 2005 issue, and the car hit 60 in 3.8 seconds, 100 in 8 flat, and ran through the quarter in 11.8 at 123.9 miles per hour.

On the track, "The steady, strong push is attended by what sounds like the MGM lion roaring in the engine room," *MT's* Frank Marcus noted. "After four such growls, the speedometer just kisses 170 miles per hour as the first braking marker flashes by."

He found the car friendlier than its Ferrari counterpart, writing, "There's none of the 'knife-edged limit handling' we criticized in the more extreme Enzo. It's even more forgiving at the limit than an Acura NSX," which was high praise indeed.

In the second half of 2006, a final batch of fifteen modified MC12s, dubbed the MC12 Corsa, was built. This track day special had a shorter nose, and the GT1's 745-horsepower V-12.

THE BIRDCAGE 75 . . .

The MC12 wasn't the only Enzo-derived Maserati. At 2005's Geneva Show, Pininfarina presented a one-off that *Car & Driver* called "one of the most sensuous automotive shapes, ever, ever, ever." The Birdcage 75 was named after Maserati's famous competition car in the early 1960s and was done to honor the design house's seventy-fifth anniversary, and the constructor's ninetieth.

The breathtaking exterior was the work of Jason Castriota. "The car was based of the MCC," the brash and immensely talented designer recalled. "We used one that had been crashed and damaged to the point where it was still fine, but they didn't want to race it any more."

He said the design's inspiration came during "August 2004 over the holiday" when he made "a sculpture of what the Birdcage would ultimately become, a teardrop suspended in a wing, ebbing and flowing." A short time later a group of Pininfarina personnel were discussing ideas for the Geneva show, and Castriota threw out the concept of reviving

From 1997 to 2005, Ferrari partially or fully owned Maserati. When Enzo production ended in 2004, Maserati got its own mid-engine hyper road car, the Enzo-based MC12. The Maserati's wheelbase was 6 inches longer than the Enzo, and the coachwork was the work of Frank Stephenson, the first design director at Ferrari's new in-house design studio.

BIRDCAGE 75 (MASERATI)

Year(s) made: 2005
Total number produced: 1
Hypercars: 1
Drivetrain: 5,998cc DOHC V-12, estimated 750 hp; six-speed manual
paddle shift
Weight (mfr): 3,300 lb
Price when new: $3,000,000

PERFORMANCE:
0–60 mph: n/a
0–100 mph: n/a
1/4 mile: n/a
Top speed: n/a
Reviewed in: *Motor Trend*, April 2006
Main competitors: A villa in you-name-where-you-want-it, fine art of
any type

the company's dream cars from the late 1960s. A design competition took place, which Castriota won.

When the team started the project in early November, they had just four months to make a running car. "The Modulo was really the inspiration for the project," he explained, referencing perhaps the most extreme Ferrari show car ever. "We did the car virtually, as it was the only way to make sure we got the engineering right in such a short time. We milled out one polystyrene model, looked at it and made two volume adjustments in literally two days. We then threw them right back into the computer and resurfaced the car from there. We really suffered through some very long nights. . . ."

At its Geneva debut, Castriota remembered the response being "so overwhelming that Andrea Pininfarina said, 'We should produce and sell this car.'" The designer and Pininfarina program manager Paolo Garella hit the road, visiting key collectors. "A number of clients thought it was fascinating," Castriota recalled, "but asked if we could do something else."

« Another Enzo derivative was the sensational Maserati Birdcage 75. Shown at Geneva in 2005, the one-off was done to celebrate Pininfarina's seventy-fifth anniversary. The exterior design done by Jason Castriota. This was the last car created under Ramaciotti's directorship at Pininfarina.

P4/5

Year(s) made: 2006
Total number produced: 1
Hypercars: 1
Drivetrain: 5,998cc DOHC V12, 650 hp @ 7,800 rpm; six-speed manual paddle shift
Weight: 2,645 lb
Price when new: Estimated $4+ million

PERFORMANCE:

0–60 mph estimated "very low 3 seconds"
0–100 mph n/a
1/4 mile: n/a
Top speed: estimated 225 mph
Road tested in: all figures from *Car & Driver*, September 2006
Main competitors: Bugatti Veyron, more fine art, Ferrari "P" car from the 1960s

. . . BRINGS ABOUT THE FERRARI P4/5

While the Birdcage remained a one-off, the roadshow led to another one-off. Investment banker and gonzo car enthusiast Jim Glickenhaus was seriously interested and had a collection that included a Ferrari 330 P3/4, the one-off Dino Competizione, and a Ford GT40 MK. IV. "All his cars are registered for the street so he can use them," Castriota said. "The problem was the MC12 wasn't legal in America, so he proposed building a Birdcage on an Enzo. That seemed a little strange to all of us, so the P4/5 grew out of that."

Glickenhaus insisted that a new unregistered Enzo serve as the P4/5's underpinnings. After a worldwide search they found the last unregistered Enzo at Ferrari of Beverly Hills (the original client decided he didn't want it), sent it to Turin, and Pininfarina's craftsmen and engineers went to work.

While Castriota had visions of doing an avant-garde Birdcage-type one-off, Glickenhaus had other ideas. "When he came," the designer said, "he wanted a replica P4, like what Ford did with the GT. While that was a beautiful car, it was really literal to the original. I wanted to push it farther, much more in the league of the Birdcage, so we had to find the happy balance.

"At the end of the day there was a client I had to satisfy. Jim was great fun to work with. He had his ideas and stuck to them. I had my ideas but luckily he was very open and willing to listen. He wasn't, 'No, I want it this way and that is it.'"

The resulting P4/5 melded the chiseled, architectural feel of the Enzo with the sensuousness of a Glickenhaus favorite, Ferrari's P4. An incredible amount of engineering went into the one-off (a large number of components were made especially for it), and the car underwent high-speed testing. Because the P4/5 weighed approximately 500 pounds less than an Enzo, it was even quicker with reported 0 to 60 times of three seconds, and a top speed of around 225 miles per hour.

The unique Ferrari broke cover at 2006's Pebble Beach Concours d'Elegance and was shown several weeks later at the Paris Auto Show. Glickenhaus was so enthused by the car that he created a second version, the P4/5 Competizione that was based on a 430 Scuderia.

In many ways, the P4/5's most enduring legacy was it sparked a rebirth of Ferrari custom coachwork. Since then Pininfarina, Zagato, and especially Ferrari's Centro Stile have made a number of one-off or very limited-production creations, though none have had the rapturous reception and impact of the P4/5.

THE FXX

This track only "Super Enzo" is the most radical Enzo of all. "The idea first came to us when we were developing the Enzo," Amedeo Felisa noted in the foreword of Phil Bachman's book *Ferrari FXX Inside Out.* "At a certain point, we asked ourselves if such extreme, powerful, and technological cars would provide our clients . . . with the driving pleasure and emotions that anyone getting behind the wheel of a Ferrari expects.

"When we decided to launch this program, we had just one goal in mind: to involve our clients as much as possible and give them an active role that would tap into the very human potential that [thus] far [had been] almost completely ignored . . . sharing with them not only the finished car, as is normally the case, but also the technological choice and the whole evolution process."

The FXX concept formed in September 2003, and Felisa surrounded himself with the right people: Antonello Coletta (head of Corse Clienti), engineer Giuseppe Petrotta, Gemignano Albergucci (technical project coordination), Giuseppe Bonollo (marketing), and Maurizio Moncalesi (production). Petrotta concisely explained the essence of the program to author (and FXX owner) Bachman as "How can we share these [advanced] technical solutions with a restricted number of customers before we apply the results to normal production?"

The first working mule was running in November 2004 and was used to sort the mechanicals. A prototype was completed in February 2005, and a sneak peak of chassis 139744 was given to a select group of Corse Clienti clients that October. The official unveiling occurred five weeks later at the Bologna Motor Show.

FXX & FXX EVOLUZIONE

Year(s) made: 2005–2007
Total number produced: 31
Hypercars: none, all track cars
Drivetrain: 6,262cc DOHC V12, 800 hp @ 8,500 rpm (FXX), 860 hp @ 9,500 rpm (Evoluzione); six-speed sequential gearbox
Weight (dry): 2,546 lb (1,155 kg)
Price when new: $1,800,000

PERFORMANCE:
    0–60 mph: 2.8–3.0 seconds, depending upon source quoted (FXX);
    2.44–2.8 seconds, depending upon source quoted (Evoluzione)
    0–100 mph: n/a
    1/4 mile: estimated n/a
    Top speed (mfr): 214 mph (FXX)
Road tested in: n/a
Main competitors: Another FXX or Evoluzione on the track

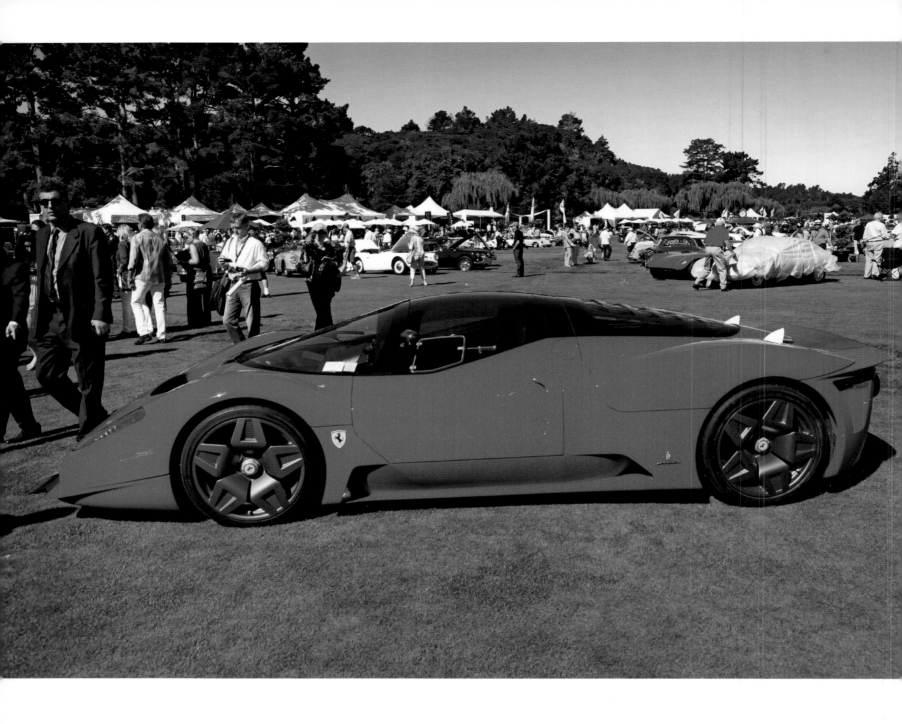

≈ The one-off Birdcage had such a raucous reception that serious consideration was given to producing a small series. One prospective client was Jim Glickenhaus, who passed but commissioned the one-off Ferrari P4/5. Also designed by Jason Catriota, it beautifully melded the Enzo's architecture with the sensuousness of the 1960s' P-car racers.

Besides having enough scoops, air intakes, and wings to make the FXX look like the perfect car to come blasting out of the Bat Cave, it shed 200-plus pounds off the production Enzo to bring the dry weight down to 1,155 kilograms (2,546 pounds). The naturally aspirated V-12 was enlarged to 6,262cc and had improved heads and pistons, a more radical cam profile for high-end power, a redesigned crankcase, and a free-flow exhaust system. This bumped power to 800 horsepower at 8,500 rpm, and torque 506 lb-ft at 5,750 rpm. Two years later, in the FXX Evoluzione version that debuted on October 28, 2007, at Mugello, power jumped to 860 at 9,500 rpm.

The six-speed sequential gearbox also received updates from the Enzo, shift times dropping from its 150 milliseconds to 80 milliseconds in the standard FXX and 60 milliseconds in the Evoluzione. Other mechanical highlights included upgraded brakes, aerodynamics, and a traction control system that now featured nine settings.

Around Fiorano, the FXX needed "less than 1:18" (*Road & Track*) or "slightly under 1:16," (Phil Bachman), the Evoluzione version dropping "approximately two seconds" off that time, according to Bachman.

So what did something that fast feel like? "It really is a weapon," Jason Barlow wrote in *CAR*'s January 2006 issue about riding shotgun as Ferrari F1 test driver Luca Badoer wailed around Fiorano. "The way the FXX generates energy is just awesome, but more awesome still is what it does with it and the way it refuses to waste it." Or, as Bader put it, "In terms of power and acceleration, even the gearbox, this is the closest thing to F1 I've driven."

While most sources state twenty-nine FXX's were made, Bachman chassis number list puts production at thirty-one. Key to the program's success was it being "track only," for this allowed Ferrari to explore technologies and solutions without the constraint of FIA or road car regulations. It also allowed the research to be underwritten in good part by select clients who could provide insightful feedback and would likely buy whatever came about from the program.

It also portended the future. In 2006, *Road & Track*'s Dennis Simanaitis asked Amedeo Felisa if the FXX represented "the Enzo II." "Maybe 90 percent of the thinking for the Enzo replacement will evolve from the FXX project," he replied. "But we also have some special ideas, some strange ideas. . . ."

⌃ The most "radical" Enzo of all was the FXX program that began in earnest in late 2004. Basically a rolling lab on wheels, the first car was shown to select clients in October 2005. Months later as the orders came in, the Enzo/ MC12 production line was busy once again. *Matt Stone*

≋ Included in the Evoluzione upgrades was bumping the FXX's 6,262cc
V-12's power from 800 to 860 horsepower at a dizzying 9,500 rpm, and
shift times dropped from 80 to 60 milliseconds.

» For a company like Ferrari, "insanely fast" is never fast enough, even on
something like the FXX, so in 2007 the FXX Evoluzione package was
introduced. "We had just one goal in mind [with the FXX program],"
Amedeo Felisa told author and FXX owner Phil Bachman, "to involve our
clients as much as possible [by] sharing with them . . . the technological
choice and the whole evolution process." Shown here (and in the other
photos) is FXX Evoluzione chassis 145749.

OPPOSITE PAGE: A no-frills interior that projected into the future.
"Maybe 90 percent of the thinking for the Enzo replacement will evolve
from the FXX project," Felisa told *Road & Track*'s Dennis Simanaitis.
Such special models are a brilliant proposition for Ferrari—clients give
immediate feedback while helping to underwrite future models they
themselves will buy.

8 BRAVE NEW WORLD
(2010–TODAY)

MELBOURNE AUSTRALIA

AUSTRALIA MELBOURNE

⌃ Formula 1's 2009 season introduced a slew of new technologies, one of the most prominent being Kenetic Energy Recovery System (KERS). The system harnessed energy created during braking into electrical energy that, when certain race conditions were met, provided an 80-horsepower boost for six-plus seconds per lap. *LAT*

SINCE 1949 AND JAGUAR'S XK120 AD, top speed had been the be-all/end-all criteria by which "fastest" was defined. Then came the F50, and no one in or outside Maranello had any clue as to how that car would completely alter the hypercar game. Its stated single-mile-per-hour increase over its predecessor waved away maximum velocity as the only measure.

With the F50, Piero Ferrari and his men had focused on what had been a key supercar tenet since the beginning: giving the driver and passenger an experience not found anywhere else. But they approached that tenet with broader objectives. When Maranello learned of the McLaren F1's existence, and its projected (and then realized) 231-mile-per-hour maximum, "It didn't change anything, because top speed was never the F50's objective," recalled an upper-management insider. "It was to create an F1 car on the road, and with the rule changes in Formula 1, this was a way of getting a V-12 into a road car."

When the rest of the hypercar pack recognized that the big dog on the porch was no longer chasing top speed, a divergence occurred: Outliers went after the highest velocity, while Ferrari and other established marques kept their formidable engineering talents focused on packing ever more advanced technology into their offerings.

That approach yielded different, broader bragging rights. Yes, the Enzo's 218-mile-per-hour maximum was well short of the McLaren's 231, but after the instruments were hooked up, the road tests were done, and everyone went home, the Ferrari was widely hailed as the greatest performance car ever.

FOLLOWING THE BREADCRUMBS

That "greatest" strategy is what Maranello pursued with its current hypercar, which would be known internally during its development as the F150. Not long after Amedeo Felisa made his "90 percent of the thinking for the Enzo replacement will evolve from the FXX project" comments in 2006, Ferrari began sprinkling clues as to what would make up its next "ultimate."

The first appeared in June 2007 during the company's sixtieth anniversary, when it presented the static Millechili model at Fiorano. Looking much like a four-fifths scale Enzo and called "arguably the most important concept car Ferrari has ever shown" by *CAR*, the name meant "1,000 kilograms." The concept was a technological showcase that forecast a concerted effort by Ferrari to diminish weight wherever possible.

"If I ask my engineers to make a 1,000-horsepower car," Piero Ferrari commented, "this is very easy for us to do for we can turbocharge one of our V-12s. But to make a 1,000-kilogram car, that is very difficult because you must closely examine everything, right down to the screws. This was the idea behind the Mille Chili."

Cutting-edge materials weren't the only approach considered. Seats would be fixed to the floor, which lowered the roofline and decreased the car's overall length, thus reducing body mass. Fixed seats also eliminated seat rail weight and electric motors to move the seats.

Other concepts included Brembo's still-under-development carbon-fiber CCM-2 brake discs, which reduced unsprung weight. Active aerodynamics were projected to improve downforce during cornering, then were altered to make the car slice through the air more cleanly, thus improving gas mileage and straight-line performance.

Such measures offered a benefit not normally associated with performance—Ferrari was showing a "green" side. And it had to, for Maranello was well aware that climate change and dwindling oil supplies were hot media topics. Indeed, so strong was public interest in climate change that *Time*'s April 3, 2006, special issue was dedicated to global warming. The cover headline stated, "Be Worried. Be Very Worried," while a bullet point noted "Earth at the Tipping Point."

Equally alarming was the rise in oil prices. In January 1999, when Enzo development was getting underway, the price was $17 a barrel on America's NYMEX, about where it had been for most of the decade. Then per-barrel cost began a steady climb. By the time the Mille Chili was presented, crude was around $70 per barrel. That 400-plus percent increase in just eight years alarmed many people, and books such as *The End of Oil*, *A Thousand Barrels A Second*, and *Twilight in the Desert* gained much traction by making the case that oil production was heading toward a tipping point when it would be outstripped by global demand.

The last thing Ferrari or any manufacturer in the high-performance industry wanted was a repeat of the vilification and near-death experience of the 1970s, so Formula 1 became a surprising vanguard to keep the series (and cutting-edge performance technologies) relevant. "New developments in F1 should be those that are directly helpful to the car industry worldwide," FIA president Max Mosley explained at a press conference in November 2006. This included instigating an engine freeze for several years and the development of "energy recovery and regenerative braking systems" that would become available in the 2009 season. "It is only by doing so that we can prevent F1 starting to be labeled as a dinosaur," Mosely said.

Ferrari's Mille Chili foretold how such systems could be applied to future road cars, where hybrid drivetrains with an electric motor would provide a blast of torque when needed. The company was also eyeing direct injection to boost horsepower and efficiency, lessons learned from previous Formula 1 seasons before the technology was banned.

"The main problem we have with consumption isn't an economic one, it's political," Amedeo Felisa said in *CAR*'s *How Ferrari Is Going Green* story. "We are working hard to

⌃ The main reason technologies such as KERS came into F1: "to prevent F1 [from] starting to be labeled as a dinosaur," FIA president Max Mosely said in November 2006. He had good reason to sound the alarm: Issues such as peak oil, global warming, and climate change were making headlines around the world.

improve our emissions but with special products like ours we can't fulfill the emissions requirements that are being talked about. The [production] numbers are just too low. We understand this is the new trend, but we . . . will not destroy the philosophy of our product. We will look at ways . . . to think in a completely different way."

HY-KERS ENTERS THE LEXICON

A "different way" sneak peek came at the 2010 Geneva Show, when Ferrari surprised everyone with a vibrant green one-off 599 GTB. Dubbed the 599 HY-KERS (Hybrid-Kenetic Energy Regeneration System), the concept car was a showcase for a hybrid powertrain that was only theoretical three years earlier.

Ferrari targeted 1 horsepower for every kilogram of weight added, and the HY-KERS system gave a 100 horsepower boost while retaining the 599's front-mounted V-12 and rear-wheel-drive powertrain. The gearbox was now a double-clutch unit and not Ferrari's traditional electro-hydraulic manual, and a high-voltage motor was connected directly to the transmission, mounted to one side. The lithium ion battery pack was placed under the seats for a low center of gravity, while mounted on the front of the engine was an electric power pack that took care of ancillaries such as air conditioning and power steering.

"Our program started in 2007 with a view to ensure that Ferrari could meet future legislation without losing its soul," technical director Roberto Fedili told Johann Lemercier in *FORZA*'s cover story on the car. "Technical evolution, government legislation, and our customer needs are thus the three factors driving this research."

In charge of the system's development was Claudio Silenzi, who was responsible for the development of Ferrari's KERS for 2009's F1 season. "Where the focus in F1 was on performance," he told Lemercier, "it has shifted here toward lowering CO_2 emissions. We transferred over the basic components, although to different specifications, as well as the software control logic. We have many ideas on how to improve the technology . . . and really so far have only scratched the surface on what can be achieved."

How such technology would actually be used was unveiled at the 2012 Beijing Motor Show, when the company displayed a very trick engine/transmission unit. "You're looking at the mid/rear engine V-12 application of Ferrari's HY-KERS hybrid system," *Car & Driver* noted of the unexpected debut. "The reason you care is because a version of this powertrain will put out at least 900 horsepower in the next Ferrari super-duper car."

The powertrain was quite similar to the experimental 599's, though the mid-engine configuration dropped the distance between the engine and transaxle to zero. A small motor sitting on top of the transmission was responsible for boosting power output and recapturing energy during deceleration, while another motor at the front of the engine worked the ancillaries.

That reveal spotlighted how the hypercar game now emphasized software and systems integration. "During braking, the electric motor acts as a generator, using kinetic energy from the negative torque to recharge the batteries," Ferrari said. "This crucial task is managed by a dedicated ECU, also F-1 derived, which not only controls the electric motor but also governs the power for the auxiliary systems. . . ."

Ferrari later stated the HY-KERS system would cut the 0-to-124-mile-per-hour (200-kilometer-per-hour) time by 10 percent, all while reducing emissions by 40 percent.

POKER GAME

Between 2010's 599 HY-KERS and the Beijing F150 powertrain debut, that brave new electronics/hybrid world had truly opened up when Porsche officially threw the first cards on the table with the debut of its plug in 918 at the 2010 Geneva Show. Six months later, Jaguar pulled the cover off its own plug in hybrid hypercar, the C-X 75. BMW would make some noise with its off-in-the-future i8, as would Honda with an Acura NSX show car.

OPPOSITE PAGE TOP LEFT: Ferrari didn't limit its Energy Recovery System research to Formula 1. Maranello's engineers saw it as a method for meeting upcoming emissions standards and boosting performance. That line of thinking was revealed at the 2010 Geneva Auto Show with the 599 HY-KERS concept. *John Lamm*

OPPOSITE PAGE TOP RIGHT: In what was a repeat performance of the 1980s' Group B/959 days, at the 2010 Geneva Show Ferrari found out it wasn't the only performance manufacturer exploring hybrid technologies. Porsche's 918 Spyder stole the show spotlight when Stuttgart made it clear the car would go into production. *John Lamm*

OPPOSITE PAGE MIDDLE LEFT: Archrival McLaren was another name that threw its hat into the hybrid-performance arena. At 2012's Paris Show it displayed what it called a "design study" of its P1 hypercar; the production model then debuted several months later at that memorable 2013 Geneva Show (shown). *LAT*

OPPOSITE PAGE MIDDLE RIGHT: Also battling for the spotlight at Geneva 2013 was Lamborghini's wild Veneno. It was a more traditional hypercar with a monstrously powerful V-12 and no electronic boost. Just three examples were going to customers, and all had been presold. *John Lamm*

OPPOSITE PAGE BOTTOM: As Porsche, McLaren, and others such as Jaguar touted their upcoming hybrid hypercars, Ferrari didn't idly sit by and let them steal the spotlight. At Paris 2012, Maranello had the F150 tub on display; here it is seen at a press briefing in Maranello. With the tub is Ferrari's technical director of GT cars, engineer Roberto Fedeli. *John Lamm*

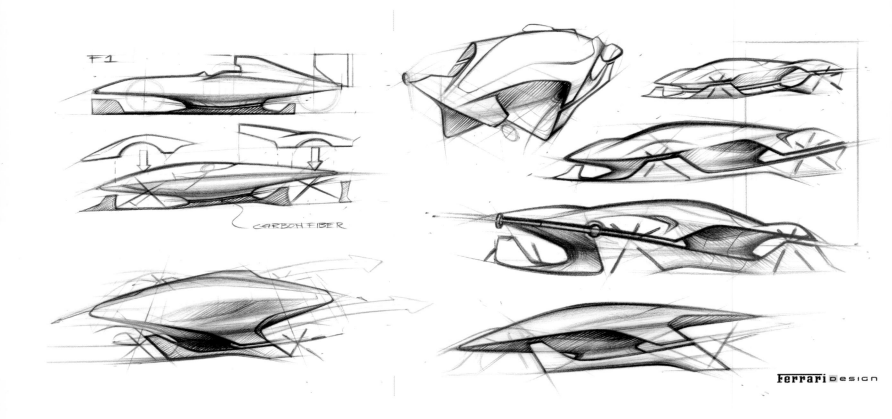

In 2002 Ferrari formed its own in-house design studio. Over the decade the studio found its footing, especially under the tutelage of its current design director, Flavio Manzoni. This concept drawing expresses key design elements in what would become known in-house as the F150. *Ferrari S.p.A.*

Following an exceptionally intense competition, for the first time since the mid-1950s Pininfarina would not be designing a Ferrari hypercar. This Ferrari Design Studio modeler is creating a scale model of the F150. *Ferrari S.p.A.*

A full-scale F150 model undergoes a comprehensive review in Ferrari's Design studio. The model has a complete interior and functioning doors and is closely scrutinized by Ferrari Chairman Luca Cordero di Montezemolo and company vice president Piero Ferrari. *Ferrari S.p.A.*

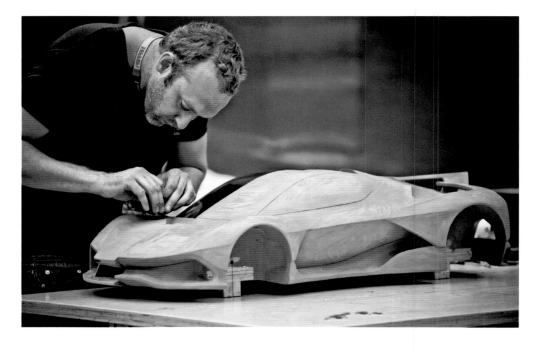

The big question remained McLaren, Ferrari's arch nemesis. In 2007, breaking stories stated that the fabled constructor of Can-Am and F1 racers had Maranello firmly in its sights by planning a full line of blisteringly fast mid-engine road cars. The twin turbo V-8-powered MP4-12C was unveiled in the second half of 2009 and went on sale in 2011. It proved formidable competition for the 458. It was quicker in a straight line, though it typically lost the closely fought comparison tests.

All these high-tech developments left everyone curious about the upcoming McLaren F1 successor. First known as the P12 and finally the P1 in production guise, it was rumored to be a hybrid with avant-garde technology, wild styling, and insane horsepower and performance numbers. At the 2012 Paris Auto Show the cover was whisked off a "design study" teaser that was said to look 95 percent like the final car. "Twenty years ago we raised the supercar performance bar with the McLaren F1," company chairman Ron Dennis said at the unveiling. "Our goal with the McLaren P1 is to redefine it once again."

"Our aim is not necessarily to be the fastest in absolute top speed," McLaren managing director Antony Sheriff seconded, "but to be the quickest and most rewarding series production road car on a circuit. It's the true test of a supercar's all-round ability and a much more important technical statement. It will be the most exciting, most capable, most technologically advanced and most dynamically accomplished supercar ever made."

Unlike the early 1990s' F50/F1 days, when Ferrari idly sat by as its new competitor (and now main adversary) told the world what they were going to do, on Ferrari's stand at that Paris show was the F150's carbon-fiber tub. "Although it was surrounded by red rocketry, showcasing Ferrari's current generation V-8 and V-12 engines, the homely but brand-spanking-new black tub was the company's real showstopper," *Car & Driver* stated. "Not only does the new mid-engine monocoque tub benefit from Ferrari's Formula 1 experience, it's being developed in conjunction with the F1 team."

Whereas McLaren remained very tight-lipped regarding the P1's mechanical specs, Rory Byrne—the architect of Michael Schumacher's championship F1 cars—openly extolled the virtues of Maranello's carbon-fiber technologies. As *C&D* noted from its interview with the legendary engineer, "The new tub uses the right kind of carbon fiber in the right thickness for every specific need throughout the chassis. Some of it is unidirectional, some of it extra-high tensile for more structural surfaces, and most all of it sandwiches between a layer of Nomex Honeycomb."

To construct that F150 tub, Ferrari used the same made-by-hand methods as on its Formula 1 cars; it was later revealed that the production tubs would be made in the Gestione Sportiva (Ferrari's F1 department). T1000, a high-strength carbon fiber with a specially formulated resin, was used for the crash structures. The hybrid drivetrain's battery tray was placed behind the seats, and used a structural layer of Kevlar. The narrow roof's spine was part of the monocoque, which added rigidity. And, as seen in the Mille Chili, the seats were built into the chassis, and the adjustable pedal box and steering wheel moved to fit the driver.

According to Ferrari, all this wizardry created a structure with 27 percent greater torsional rigidity, 22 percent better beam rigidity, and 20 percent less weight than the Enzo's chassis. Most impressive, this was accomplished in a period with more stringent crash standards.

DESIGN CHANGES

The F150 marked the first time since the late 1950s that Pininfarina would not define what Ferrari's next hypercar would look like.

The seeds of Maranello's own in-house design studio's prominence were sown in 2002 when Ferrari appointed former Ford and BMW designer Frank Stephenson as director of its newly formed Ferrari-Maserati Design division. He worked closely with

⤊ What makes the launch of Maranello's latest hypercar different from every other manufacturer? The lengths to which normally jaded automotive journalists will go to find any place they can to witness the reveal. *LAT*

⤊ At the F150's 2013 Geneva Show reveal, Montezemolo extols the virtues of Ferrari's latest hypercar. "Usually, industrialists are cold and sharp," Montezemolo's friend Paolo Borgomanero told the *Wall Street Journal's* Joshua Levine. "That's not Luca. When he's happy, you know about it, and when he's not happy, you know about it." Here, he is quite happy. *LAT*

Pininfarina on the Maserati Quattroporte, and Ferrari's 612 Scaglietti and 430, and was largely responsible for the appearance of the MC12. But Stephenson and Ferrari never designed a car entirely in-house. When he was transferred to head Fiat's Design Center in 2005, Ferrari hired Donato Coco from Citroen. Coco held the position until late 2009 when he was hired by Lotus.

Replacing him was Flavio Manzoni. Then in his mid-forties, Manzoni was born in Sardinia and had graduated from college with a degree in architecture with an emphasis in industrial design. He entered the auto industry in 1993 with Lancia, and served a stint at Spain's SEAT before returning to Lancia as its design director in 2001—a role that expanded to include Fiat. In 2006 he moved to Volkswagen and became its director of creative design before joining Maranello in January 2010.

With his arrival, Ferrari's in-house design plans became much more ambitious. One likely reason was the prolonged reversal of fortunes Pininfarina had experienced after Sergio Pininfarina's retirement in 2001. While the company's design arm remained at the forefront of the auto industry, the manufacture of car bodies by independents had all but dried up by the middle of the decade. That unused capacity hit cash flow hard, causing Pininfarina to stop paying a dividend to shareholders in 2005. The stock price, which had been floating between 20 and 30 euros for much of the decade, declined precipitously in late 2007 with the onset of the Great Recession. Then Andrea Pininfarina, Sergio's eldest son who succeeded him as the company's head, was killed in July 2008 while riding a Vespa to work. A year later, Reuters reported that the Pininfarina family was having Italy's Banca Leonardo sell their majority stake.

Along with Manzoni's hiring came additional personnel, many of them specialists in areas such as aerodynamics. More resources were dedicated to the design department, including constructing a virtual-reality room where a design team, engineers, executives, and clients can see upcoming and under-consideration models and prototypes in full-size 3D from any angle through the use of sophisticated computer graphics and projection systems.

In mid-2010 Manzoni gave his designers and Pininfarina a briefing on the F150. "We said we had to create something, a kind of milestone," he told *Autocar*'s Steve Sutcliffe. "Mr. Montezemolo wanted something really strong, original, futuristic, and not conservative. We tried to learn a lot about Ferrari DNA, the Ferrari tradition, but I didn't want any kind of retro feeling . . . (anything) coming from the past and simply translated into a modern language."

The teams in both design houses created a number of renderings. Manzoni later commented to *Classic Driver*'s Jan Baedeker, "At the beginning of a new project I create a competition . . . which gives excellent motivation to the designers. As soon as we reach a more mature phase of the project, we try to converge the best ideas into one model. It is a synergy."

The synergy's first step happened in October when ten 1:4 scale models were presented at a meeting Manzoni said was nicknamed "involucro" ("covering" or "wrapper" in English). Also at the presentation was a 1:1 "technical model developed by the pre-engineering department. It was a functional model representing all the needs regarding aerodynamics and so on, and the original idea was to keep the car very small. But it wasn't very nice to look at . . . a bit brutal, a little bit primitive. Once we understood the complexities of the aerodynamics—for instance, the pressure to create airflow toward the side of the car and the radiators—we could move forward."

The original ten proposals (of which one was so intriguing that as of this writing, it has not been seen in public) led to five main themes. These were made into 1:1 scale models and, at a presentation in May 2011, three were chosen (two by Ferrari's Design Center, one from Pininfarina).

"Two concepts were then selected," Manzoni noted in the book *LaFerrari Dynamic Art*. "After a round of adjustments, one of the two Design Center's proposals was finally chosen. . . . The following stage involved long, demanding development, making adjustments to the model, perfecting it in every aspect: aesthetics, aerodynamic performance, and technical feasibility.

"The creation of the in-house virtual room, along with the construction of surface plates for the 1:1-scale clay models was crucial [to] adopting a modern, avant-garde, and rapid methodology. It was possible to reduce the length of every feedback loop by milling the clay at low speeds directly on the surface plates and scanning them, thus creating a process that was synchronized in real time with the engineering department."

BRAVE NEW WORLD

The design team needed such close collaboration. As Ferrari Technical Director Roberto Fedeli, who graduated with a degree in aerospace engineering, noted in *LaFerrari Dynamic Art*, "The car must be highly efficient at both high speeds on straights and on twisty roads. This led to the decision to research two different but complimentary aerodynamic decisions: HDF (high down force) and LDF (low down force). These were realized through the introduction of active aerodynamic devices both on the front and rear of the car that, changing both the interior and exterior airflows, changes downforce and drag."

Thanks to the active rear spoiler, a front guide vane, flaps in the front and rear diffusers, and more, the F150's Fiorano lap time dropped five seconds when compared to the Enzo. "This landmark result was obtained through an uncompromising approach to the conceptual design," Fedeli explained in *Dynamic Art*. "All the engineering characteristics of the individual components which make up the vehicle have been thought out and developed to be completely integrated: not just with each other, but with the control logics that manage those very components."

⩔ The view everyone will see of Ferrari's latest and greatest. When *Motor Trend* put instruments on a McLaren P1, it hit 60 in 2.6 seconds, 100 in 4.7, and the quarter mile in 9.8 seconds at 148.9 miles per hour. LaFerrari will likely be faster, as it weighs less and has 50 more horsepower, all playing to the tune of that V-12 wail. *LAT*

≈ What most everyone overlooks is that not only is Ferrari *the* benchmark performance car manufacturer, but it's also a fiercely independent proprietary software firm on par with Apple. "The F150 hybrid integration is the hardest engineering we have ever done," technical director Roberto Fedeli told the *Wall Street Journal*'s Dan Neil. *LAT*

» The helm of the starship *Enterprise*—all thrills and no frills. If something didn't serve a function, or if it added extra weight, Ferrari left it out. Indeed, the pedals and steering wheel move because the driver sits on the floor with custom formed cushions, which eliminated the weight and height of frame rails. It also made the roofline lower. *LAT*

The car's dynamic control systems included E-diff electronic control, eF1-Trac traction control (which came from F1 and was also integrated with the HY-KERS system); SCM damping control with "lightweight shock absorbers in an aluminum-nickel alloy as well as new software algorithms"; ESC stability control for "the car's lateral dynamics"; and ABS with "a new logic . . . to regenerate energy during braking even when braking at the limit."

"The integration of systems is not a simple sum of functions," Fedeli wrote, "but a work of art, where various electric systems intervene at the right moment and at the right level, to effortlessly achieve levels of performance that were previously unheard of. All this is coherent with what Aristotle declared: 'The whole is greater than the sum of the parts.'"

"LAFERRARI" MEANS EXCELLENCE

And so the decades-long story of Maranello's ultimate performance cars returns to where our story began—the 2013 Geneva Auto Show. "This is proper, once-in-a-generation stuff," EVO noted in its "Modern Warfare" show coverage. "It's rare that we're inundated with new top-level supercars, but that's exactly what's about to happen."

When the cover came off the F150 show car (chassis 194527), the automotive world saw a near-perfect blending of classic Ferrari beauty with advanced aerodynamics. "We chose to call this model LaFerrari because it is the maximum expression of what defines our company—excellence," Luca Montezemolo explained at the unveiling. "Excellence in terms of technological innovation, performance, visionary styling, and the sheer thrill of driving. Aimed at our collectors, this is truly an extraordinary car that encompasses advanced solutions that, in the future, will find their way onto the rest of the range, and it represents the benchmark for the entire auto industry. LaFerrari is the finest expression of our company's unique, unparalleled engineering and design know how, including that acquired in Formula 1."

In addition to aforementioned technologies, highlights included a seven-speed paddle shift twin-clutch gearbox that Autocar said "could swap gears faster than you can think,"

LAFERRARI

Year(s) made: 2014–
Total number produced: 499
Hypercars: all
Drivetrain: 6,262cc DOHC V-12 plus electric motors, 963 hp @ 9,000 rpm; seven-speed dual clutch automatic transmission
Weight: 2,761 lb
Price when new: $1.35 milliion

PERFORMANCE:
0–60 mph (mfr): sub 3.0 sec
0–100 mph: n/a
1/4 mile: n/a
Top speed: 217 mph
Road tested in: none at the time of publication
Main competitors: McLaren P1, Porsche 918, Pagani Huayra, Bugatti Veyron Super Sport

and a suspension with aluminum double wishbones, coil springs over electronic adaptive shocks, and an anti-roll bar. Power output was listed at 950 to 963 horsepower at 9,000 rpm, with 790 and 516 lb-ft of torque coming from the 8,282cc DOHC V-12. The electric motor provided the other 160–173 horsepower, and an additional 199 lb-ft of torque.

While no instrumented road tests of any kind had been completed by the time of this book's publication, the quoted performance is formidable: 0–100 kilometers per hour (62 miles per hour) in "under 3 seconds," and 0–300 kilometers per hour (186 miles per hour) in 15.5 seconds, faster than either the McLaren P1 or Porsche 918. Regarding maximum velocity, Ferrari flatly states, "We simply don't care about top speed."

While those numbers are indeed impressive, perhaps most eye-opening is the car's weight. Ferrari lists it at 1,255 kilograms (2,761 pounds) dry, with the HY-KERS system accounting for 140 kilograms (308 pounds). Without the HY-KERS, the car's dry weight would be a featherweight 1,115 kilograms (2,453 pounds). Weight is something that no computer system can fake, so the car's responsiveness should be extraordinary.

About the Geneva debut, *Autoweek* named LaFerrari its "Best in Show," stating "Consider the Enzo replaced. It took nearly a decade, but it was worth the wait."

"LaFerrari is easily the most memorable car at the show," observed *Autoweek*'s news editor Greg Migliore. "Years from now, we'll always remember how we were taken with the looks and superlative power. . . . From the raw performance to the striking appearance, everything about LaFerrari is breathtaking."

« "LaFerrari joins a long and impressive lineage of Ferrari supercars: the 288 GTO . . . F40 . . . F50 . . . and Enzo," the UK's respected veteran journalist Andrew English noted in his driving impressions for the *Telegraph*. "I've driven them all and LaFerrari is by far and away the fastest. . . . [It's] the first car I've driven where the straights are scarier than the corners." *LAT*

FIORANO LAP TIMES

Ferrari's 1.86-mile-long test track is located approximately 1 mile away from the factory in the town of Fiorano. The track was inaugurated in 1972, and the course has undergone several changes over the years, the last major alterations taking place in 1996.

Fiorano lap times are the metric Ferrari uses to show its cars' increase in performance. Times of the hypercars and derivatives are:

288 GTO—1:36
F40—1:29.6
F50—1:27
Enzo—1:24.9
LaFerrari—1:19
F40 LM—1:18
FXX—1:16.2
Maserati MC12 GT1—1:13
333 SP—1:11.9
F50 GT—estimated 1:08.4–1:10.4*

*F50 GT times based on recollections of people working at the company at the time and stating it was 1.5–3.5 seconds faster than the 333.

Epilogue

WHERE DO WE GO FROM HERE?

Just weeks before this book went to press, Maranello held the first LaFerrari driving impressions at Fiorano and on the stupendous roads that weave their way through the surrounding hills. Every tester heaped superlatives on the machine, with *Car and Driver*'s Eddie Alderman summing it up best when he noted, "We might have reached the point where cars are getting so outrageously fast and so seriously powerful that they have outstripped our ability to invent new ways to describe them."

No doubt we have entered a new era on what defines "fast," and those colleagues who tried to sum up the car's character in a few words said the LaFerrari felt like a "1,000 horsepower 458." How times have changed—and not just in the civility and tractability of these ultimate performers.

Until about 2005, driving an exotic (and especially something from Italy) was like going to a fine restaurant and enjoying an unforgettable multicourse meal, one where a gourmet chef infused a unique and overwhelming mélange of obvious and subtle flavors in each course. For instance, a Ferrari 275 GTB offered a specific "taste" that could only be found in that model. It was quite distinct from the 250s that preceded it and just as different from its Daytona successor, which was equally different from the 550 Maranello that came much later.

In such cars, numbers were important but not the overriding facet of what made these machines so great. They actively talked to you, each with its own individual language and inflection, enchanting and seducing you by engaging all your senses all the time, even while sitting at idle. The mechanical symphony from the engine compartment, the exhaust burble, the minute vibration emanating from that billet shift knob and through the pedals, even from the chassis and firewall, were all subtle forms of conversation that said *I'm alive; are you ready to GO*?

Back then, the *N* and *V* in NVH (noise, vibration, harshness) were finely tuned character traits, something to relish, not revile, and certainly something not to mute because Japanese manufacturers (and then everyone else) decided they needed to relentlessly pursue a nebulous goal called "perfection." The symphony of those subtle mechanical "spices" is what made each automotive delicacy so succulent and vivid; so much so that you yearned to try it again and again or to sample something else prepared by the genius and his staff lurking in the kitchen. You savored the experience long after it was over—days, months, and even years with the greats.

Enzo Ferrari and his successors have been the auto industry's master chefs since 1947, creating a menu of unparalleled depth. A Ferrari-owning uncle introduced me to the marque in the mid-1970s; since then I've have been so very fortunate to drive so many different Ferraris that I can't say with certainty how many models it was! From 166s up to Maranello's latest, I've sampled many a Ferrari, some so memorable that it was like going on a fabulous first date with a beautiful, enchanting companion, hoping the day would never end.

CARE FOR SOME VANILLA?

Unfortunately, today's hypercars are no longer like individual meals and memorable first dates and more akin to hot dog eating contests, where the singular purpose isn't to savor anything but hoover down as much tarmac as possible in the shortest amount of time. This narrow conviction that superior numbers offer the driver and passenger a more rewarding experience has utterly corrupted driving pleasure and diminished marque differentiation, reminding me of what Aldous Huxley noted in *Brave New World*: "one believes things because one has been conditioned to believe them."

This disappearance of communicative characteristics, along with the sole focus on numbers, has made it nearly impossible to discern which manufacturer made what. If you take a blindfolded passenger out in something from the digital age, the only real difference they will notice are the g-forces felt and perhaps the exhaust note. For the most part, the cars simply pin the driver and passenger in their seats when asked and compensate for any missing skills with a variety of driving aids. No longer is there any dialogue or interaction to engage the occupants, let alone give the car a specific or unique identity.

Undoubtedly the leading cause of change has been the packing of more and more terabyte technology into cars. Computing power is what the creators emphasize today, as Maranello's own lovely book, *LaFerrari Dynamic Art*, highlights. While scouring the text for more insight into the title's technological marvel, I noticed only a single instance of a word or phrase pertaining to driver involvement and experience, while words or phrases such as *systems*, *software*, *active*, and yes, *technology* was at least 45.

Shouldn't it be the other way around? Or at least a bit more balanced?

THE HYPERCAR PARADOX

During time spent with Lorenzo Ramaciotti while researching this book, he spoke of a day at Balocco, the Fiat group's test track near Milan, where company executives brought their latest vehicles for everyone to try. Driving the LaFerrari was former F1 star Giancarlo Fisichella, who utterly astounded every passenger with the car's mind-boggling poise, acceleration, cornering, and braking.

After their LaFerrari laps, Fisichella's passengers had a similar reaction: the car was so far removed from their driving capabilities, and thus their reality, that they might as well have been passengers onboard the space shuttle. Yet, when they came away from time in Alfa's 4C, all they could talk about was what a great experience it was because they were totally immersed in the joy of driving.

My point is that with most software-era supercars, it feels like you've sat down not at a fine Steinway but rather a player piano where the machine does all the work for you. There's little conversation or connection with the car; you're simply along for the ride, moving your hands and fingers to mimic the performance, not feeling at all like an active participant.

Interestingly, such engineering prowess may only go so far. In *CAR's* November 2012 issue, Ben Barry extolled the virtues of Aston Martin's new Vanquish, then noted that Ferrari's F12 would utterly obliterate it in every measurable way. But then he went on to wonder if that really mattered, telling the story of how the test driver for "a well-known independent supercar maker" had led "his supercar-driving clients down a twisty road and had outrun them in—and he was absolutely serious—a transit minibus."

This inability to exploit even a fraction of a supercar's potential is something I call the Hypercar Paradox, which simply states that *as technology increases, usability and engagement decline.* Such searing, non-involving speed is so far beyond the reaches of 99 percent of hypercar owners that only by renting an airstrip or racetrack can they truly feel it and thus get an inkling of their car's character.

G-forces are now the last element that separates a good driving simulator from a software-era performance car, as the decades-long supercar tenet of offering an experience not found anywhere else has been sacrificed on the altar of lower times and higher speeds. Could this mean we are on the verge of a new "twilight of the goddesses," to reuse Jean-Francis Held's brilliant expression from 1972's *Automobile Year*? With distinct personalities and diverse characters all but gone, the hypercar can no longer be considered "the closest thing we will ever create to something that is alive," to use the words of Jaguar founder Sir William Lyons.

Instead they have become four-wheeled iPhones and iPads. When it's only about the hype of numbers, rather than a unique creation that delivers a haunting, lingering experience, the car becomes old news pretty quickly and drivers move on. What would compel you to use it, let alone cherish it? Autonomous cars are on the horizon, and there are other engaging activities to occupy your time. Supercars devoid of character will simply be another device or appliance, a fast mode of transportation and *not* that unparalleled experience they were for decades—a mesmerizing, talkative, unforgettable form of entertainment.

HOLDING OUT HOPE

Will Ferrari buck this trend by infusing future models with truly unique personalities and involvement at any speed? I certainly hope so, for as one colleague who has tried the P1, 918, and LaFerrari confided, these cars have become so fast as to be almost painful to drive, thus eliminating any real joy in using them.

Maranello has whipped up more unique dishes than anyone else by a substantial margin and forged influential new paths in the past: witness paddleshift transmissions and the F50's changing of the hypercar "top speed" game. As the undisputed market leader, if Ferrari does it, all the other "kitchens" will very likely follow.

After all these years, an alloy-bodied 250 SWB and Spyder California and an alloy-bodied 275 GTB/4 remain the most delicious and engaging road cars I've tested, so I'm crossing my fingers that Ferrari (and other manufacturers) will begin giving the driver-machine conversation equal footing to technology. If Maranello's future models perform as projected and chat away like those three luscious classics, well, they will definitely be all-time greats, *the* cars to keep forever and further tie you to the marque, regardless of what new forms of entertainment are conceived, or what the competition does.

Bibliography

Alfieri, Bruno, ed. *Ferrari America, Superamerica, Superfast*. Milan: Automobilia S.r.l., 1998.

———. *Pininfarina Ferrari 50 Renderings*. Milano: Automobilia, 1997.

Alfieri, Ippolito. *Ferrari F50*. Milano: Automobilia, 1996.

Anderloni, Carlo, Felice Bianchi, and Angelo Tito Anselmi. *Touring Superleggera*. Rome: Edizioni di Autocritica, 1983.

Anselmi, Angelo Tito, ed. *Carrozzeria Italiana: Advancing the Art and Science of Automobile Design*. Milan: Automobilia, 1980.

———. *Le Ferrari di Pininfarina*. Milan: Edizioni Grafiche Mazzuchelli, 1988.

———. *Tipo 166: The Original Sports Ferrari*. Sparkford: Haynes Publishing Group, 1985.

Anselmi, Angelo Tito, and Marcel Massini. *Making a Difference*. Milan: Le Edizioni dell'Opificio, 2006.

Bachman, Philip. *Ferrari FXX Inside Out*. Johnson City: Philip Bachman, 2012.

Bailey, Stephen. *La Dolce Vita*. London: Fiell Publishing Limited, 2011.

Bamsey, Ian. *Ferrari 312 & 512 Sports Racing Cars*. Sparkford: Haynes Publishing Group, 1986.

Barnes, John. *The Automotive Photography of Pete Coltrin*. Scarsdale, John W. Barnes Jr. Publishing Inc., 1978.

Bartz, Matthias. *Dino Compendium*. Kronberg: Verlag Matthais Bartz, 2011.

Beehl, Nathan. *Ferrari Berlinetta Boxer*. Bedsfordshire: Fiorano Publishing, 2007.

Bersani, Alberto, and Augusto Costantino. *Il Salone Dell'Automobile*. Torino: Daniela Piazza Editore, 1984.

Bertelli, Giovanna. "Dolce Italia." New York: Rizzoli International Publications, Inc., 2005.

Bluemel, Keith, and Jess G. Pouret. *Ferrari 250 GTO*. Devon: Bay View Press, 1998.

———. Ferrari 250 GTO. Bideford: Bay View Books, 1998.

Borgeson, Griff. *The Alfa Romeo Tradition*. Sparkford: Haynes Publishing Group, 1990.

Braden, Pat, and Roush, Gerald. *The Ferrari 365 GTB/4 Daytona*. London: Osprey Publishing Limited, 1982.

Briggs, Ian. *Endurance Racing 1982-1991*. London: Osprey Publishing, 1991.

Casucci, Piero. *Chiti Grand Prix*. Milano: Automobilia, 1987.

———. *Enzo Ferrari: 50 Years of Motoring*. New York: Greenwich House, 1982.

Cimarosti, Adriano. *Carrera Panamericana Mexico*. Milan: Automobilia, 1987.

Colombo, Gioachino. *Origins of the Ferrari Legend*. Yeovil: Haynes Publishing Group, 1987.

Coltrin, Peter, and Jean-Francois Marchet. *Lamborghini Miura*. London: Osprey Publishing Ltd., 1982.

Cornil, Etienne. *Ferrari by Pininfarina*. Milan: Giorgio Nada Editore, 1998.

Curami, Andrea. *Lancia Stratos Thirty Years Later*. Vimodrome: Giorgio Nada Editore, 2003.

Dieudonne, Pierre, and Jean Sage. *Ferrari F40 LM*. Vaduz: Cavalleria S.A., 1994.

Dregni, Michael. *Inside Ferrari*. Osceola, WI: Motorbooks International, 1990.

Drexel, John, ed. *The Facts on File Encyclopedia of the 20th Century*. New York: Facts on File, Inc., 1991.

Eaton, Godfry. *The Complete Ferrari*. London: Cadogan Books Ltd., 1986.

Ferrari, Enzo. *Una Vita Per l'Automobile*. San Lazzaro di Savena: Conti Editore, 1998.

———. *The Enzo Ferrari Memoirs*. Hamish Hamilton Ltd, London, 1963.

Fitzgerald, Warren W., Richard F. Merritt, and Jonathan Thompson. *Ferrari: The Sports and Gran Turismo Cars*. New York: W.W. Norton & Company, 1979.

Forghieri, Mauro, and Daniele Buzzonelli. *Forghieri on Ferrari*. Vimodrone: Giorgio Nada Editore, 2013.

Fowler, Steve, and Hugo Andreae, eds. *Ferrari Enzo*. Peterborough: Haymarket Magazines, Ltd, 2002.

Franzini, Davide. *Ferrari 333 SP*. Milamo: SolariS Progetti Editoriali, 1998.

Frere, Paul. *Porsche 911 Story*. Sparkford: Patrick Stephens Limited, 2000.

Froissart, Lionel. *Ferrari Pininfarina*. New York: Universe Publishing, 1997.

Gardner, Richard N. *Mission Italy*. Lanham: Bowman & Littlefield Publishers, Inc., 2005.

Gauld, Graham. *Modena Racing Memories*. Osceola, WI: MBI Publishing Company, 1999.

Gentili, Moreno, ed. *LaFerrari Dynamic Art*. Milano: Skira Editore S.p.A., 2013.

Ginsborg, Paul. *A History of Contemporary Italy*. New York: Palagrave MacMillan, 2003.

Ghini, Antonio, ed. *Ferrari 1947-1997*. Milan: Giorgio Nada Editore, 1997.

Godfry, John. *Ferrari Dino SPs*. Avon: Patrick Stephens Limited, 1990.

Goodfellow, Winston. *Giotto Bizzarrini: A Technician Devoted to Motor Racing*. Milan: Giorgio Nada Editore, 2004.

———. *ISORIVOLTA: The Men, The Machines*. Milan: Giorgio Nada Editore, 1995.

———. *Italian Sports Cars*. Osceola, WI: MBI Publishing Company, 2000.

———. *Ferrari Road & Racing*. Lincolnwood: Publications International Ltd, 2005.

Goodfellow, Winston, and Beverly Rae Kimes. *Speed, Style and Beauty*. Boston: MFA Publications, 2005.

Grayson, Stan, ed. *Ferrari: The Man, The Machines*. Princeton: Princeton Publishing Inc., 1975.

Harvey, Mike, and Mark Harrop, eds. *Ferrari F50: The Ultimate Ferrari*. Teddington: Haymarket Motoring Publications, 1995.

Henry, Alan. *Ferrari: The Grand Prix Cars*. Richmond: Hazelton Publishing, 1984.

———. *Ferrari Prototype Era*. Phoenix: David Bull Publishing, 2004.

———. *Ferrari: The Battle for Survival*. Sparkford: Patrick Stephens Ltd., 1996.

Higham, Peter. *The Guinness Guide to International Motor Racing*. Enfield: Guinness Publishing Ltd., 1995.

Hill, Phil. *A Champion's View*. Ferring: Dalton Watson Fine Books Ltd., 2004.

Hodges, David. *Ford GT40*. Croydon: Motor Racing Publication LTD, 1998.

Hoffman, Paul. *That Fine Italian Hand*. New York: Harry Holt & Co., 1990.

Hope-Frost, Henry, and John Davenport. *The Complete Book of the World Rally Championship*. St. Paul: Motorbooks International, 2004.

Hughes, Mark. *Ferrari F40*. London: Salamander Books, 1990.

Klein, Reinhard, and John Davenport. *Group 4 From Stratos to Quattro*. Koln: McLein Publishing, 2011.

———. *Group B: The Rise and Fall of Rallying's Wildest Cars*. Koln: McLein Publishing, 2012.

Klementaski, Louis. *Klementaski & Ferrari*. Milan: Automobilia, 1991.

Lamm, John. *Velocity Supercar Revolution*. St Paul: MBI Publishing, 2006.

Leffingwell, Randy. *Porsche 911: 50 Years*. Minneapolis: Motorbooks, 2013.

———. *Porsche 911 Perfection by Design*. St. Paul: MBI Publishing Company, 2005.

Levi, Carlo. *Fleeting Rome*. Chichester: John Wiley & Sons, 2005.

Lewandowski, Jurgen. *Ferrari GTO*. Munich: Sudwest Verlag GmbH & Co., 1987.

Ludvigsen, Karl. *Ferrari by Mailander*. Ferring: Dalton Watson Fine Books, 2005.

———. *Porsche: Excellence Was Expected*. Cambridge: Bentley Publishers, 2008.

———. *Ferrari: 60 Years of Technical Innovation*. Maranello: Ferrari S.p.A., 2007

Lynch, Michael, William Edsgar, and Ron Parravano. *American Sports Car Racing in the 1950s*. Osceola, WI: MBI Publishing Company, 1998.

Manton, David. *Enzo Ferrari's Secret War*. Auckland: Bridgehampton Publishing Ltd., 2011.

Marzotto, Giannino. *Red Arrows Ferraris at the Mille Miglia*. Milan: Giorgio Nada Editore, 2001.

Massini, Marcel. *Ferrari 250 LM*. London: Osprey Publishing Limited, 1983.

———. *Ferrari by Vignale*. Milan: Giorgio Nada Editore, 1993.

McCarthy, Patrick. *Italy Since 1945*. New York: Oxford University Press, 2000.

Merritt, Richard F. *Ferrari Brochures and Sales Literature 1946-1967—A Source Book*. Scarsdale: John W. Barnes, Jr. Publishing, Inc., 1976.

Moity, Christian, Jean-Marc Teissedre, and Alain Bienvenu. *24 Heures du Mans 1923-1992*. Editiond d'Art J.P. Bathelemy, 1992.

Moretti, Valerio. *Ghia*. Milamo: Automobilia, 1991.

Mosquera, Carlos, and Enriquem Coma-Cros. *Ricart-Pegaso*. Barcelona: Arcris Ediciones.

Nichols, Mel. *Ferrari Berlinetta Boxer 365 & 512 Series*. London: Osprey Publishing Limited, 1979.

———. *And the Revs Keep Rising*. Sparkford: Haynes Publishing Group, 2013.

Nowak, Stanley. *Ferrari on the Road*. Belton: Dalton Watson Fine Books, 1993.

Nye, Doug. *Dino: The Little Ferrari*. St. Paul: Motorbooks International, 2004.

———. *Ferrari 375 Plus*. Vaduz: Cavaelleria S.A., 1994.

Nye, Dough, Ron Dennis, and Gordon Murray. *Driving Ambition*. London: Virgin Books, 1990.

Orefici, Oscar. *Carlo Chiti: Sinfonia ruggente*. Rome: Autocritica Edizioni, 1991.

Orsini, Luigi, and Franco Zagari. *The Scuderia Ferrari*, Osprey Publishing Ltd, London, 1981.

Pascal, Dominique. *Ferraris at Le Mans*. Sparkford: Haynes Publishing Group, 1986.

Pininfarina, Battista. *Born with the Automobile*. Milan: Automobilia, 1993.

Porter Philip. *Jaguar Sports Racing Cars*. Bideford: Bay View Books, 1995.

Pourret, Jess G. *The Ferrari Legend: 250 GT Competition*. Scarsdale: John W. Barnes, Jr. Publishing, Inc., 1977.

Pritchard, Anthony. *Scarlet Passion Ferrari's Famed Sports Prototype and Competition Sports Cars 1962-73*. Sparkford: Haynes Publishing Group, 2004.

———. *Ferrari Men from Maranello*. Sparkford, Haynes Publishing Group, 2009.

———. *Ford vs. Ferrari*. Marina del Rey: Zuma Marketing, 1984.

Prunet, Antione. *The Ferrari Legend: The Road Cars*. New York: W.W. Norton & Company, 1981.

———. *Ferrari Sports Racing and Prototypes Competition Cars*. New York: W.W. Norton & Company, 1978.

———. *Pininfarina Art and Industry 1930-2000*. Vimodrone: Giorgio Nada Editore, 2000.

Purdy, Ken. *Ken Purdy's Book of Automobiles*. Chicago: Playboy Press, 1972.

Quinn, Edward. *Riviera Cocktail*. Kempen: teNeues Publishing Group, 2011.

———. *Stars and Cars of the 1950s*. Kempen: teNeues Publishing Group, 2011.

Rancati, Gino. *Ferrari A Memory*. Osceola, WI: Motorbooks International, 1989.

Ridgley, Dyke W. *Ferrari 410 Superamerica Series III*. Dyke Ridgley Publishing, 1983.

Rogliatti, Gianni. *Maranello Ferrari e La Sua Gente*. Maranello: Edizioni del Puntografico.

———. *Ferrari 125*. Maranello: Edizioni del Puntografico.

Siebert, David. *Ferrari 333 SP*. Vaduz: Cavalleria S.A., 1996.

Shelby, Carroll. *The Cobra Story*. New York: Trident Press, 1965.

Shoen, Michael. *The Cobra-Ferrari Wars*. Vancouver: CFW, 1987.

Skilleter, Paul. *Jaguar: The Sporting Heritage*. London: Virgin Publishing Group, 2000.

Small, Steve. *The Guinness Complete Grand Prix Who's Who*. Enfield: Guinness Publishing, 1994.

Spain, Ronnie. *GT40*. St. Paul: Motorbooks International, 2003.

Thompson, Jonathan. *Boxer the Ferrari Flat-12 racing and GT Cars*. Costa Mesa: The Newport Press, 1981.

———. *Ferrari Albums 1, 2 & 3*. Northbrook: The Color Market, 1981.

Trow, Nigel. *Lancia Stratos World Champion Rally Car*. London: Osprey Publishing Ltd., 1990.

———. *Lancia Racing*. London: Ospfrey Publishing, Ltd., 1987.

Varisco, Franco. *815 The Genesis of Ferrari*. Hyde Park Group, 1990.

Wimpffen, Jano. *Time and Two Seats Five Decades of Long Distance Motor Racing*. Redmond: Motorsports Research Group, 1999.

Zana, Aldo, ed. *Enzo Ferrari History and Glory of Modena Motoring*. Modena: Museo Casa Enzo Ferrari, 2013.

Periodicals and Annuals: *Annual Automobile Review/Automobile Year, Automobile Quarterly, Autocar, Autosport, CAR, Cavallino, Car & Driver, EVO, FORZA, Motor, Motor Trend, Quattroroute, Road & Track, Sports Car Illustrated, Sports Car International, Style Auto, Time, World Car Catalogue, World Cars.* The Brooklands Books series of magazine article reprints were a great help as well.

Websites: There is much erroneous information on the Internet. However, three websites trustworthy enough to use as a start to research or to look to for collaboration are Barchetta.cc, Supercars.net, and Wikipedia.org.

Index

Page numbers in bold italics indicate an item that appears in a photograph or caption.

Agnelli, Gianni, **73**, 73–74, 82, 88, 190
Albergucci, Gemignano, 215
Alesi, Jean, 142
Alfa Romeo, 17, 21, 23, 31, 37
    8C, 31
    158, 31
    159, 37
    164, 127
    P3, **17**, **18–19**
    Spider, 94
Anderloni, Carlo Felice Bianchi, 28–29, **28**, 68
Anderloni, Felice Bianchi, 21, 27
Artioli, Romano, 154
Ascari, Antonio, 21
Aston Martin Vanquish, 235
ATS, 65
Audi, 157
Auto Age, 49
Autocar, 29, 127, 135, 163, 206, 233
Automobile, 128
Automobile Quarterly, 17
Automobile Year, 88
Autosport, 31, 35
Autoweek, 233

B. Engineering, 98
Bachman, Phil, 217
Badoer, Luca, 217
Balestre, Jean-Marie, 140
Bao Dai, 46
Barlow, Jason, 217
Barnard, John, 104, 161
Barry, Ben, 235
Bell, Roger, 110
Bellei, Angelo, 81, 84, 86
Bellentani, Vittorio, 21
Bentley R-Type Continental, 53
Benuzzi, Dario, **124**, 142, 163, **164–165**, 203
Berger, Gerhard, 131
Bertocchi, Aurelio, 87
Biondetti, Clemente, **27**
Bizzarrini, Giotto, 13, 65
Bizzarrini GT Strada, 74
Bluemel, Keith, 61
BMW i8, 225
BMW M1, 116
Boano, Mario, 35
Bonollo, Giuseppe, 215
Borgeson, Griff, 17
Bott, Helmuth, 120
Brovarone, Aldo, 71, 73, **73**, **123**, 126
Brun, Walter, 131
Bugatti, 98, 154

Bugatti EB110, **99**, 154, 163
Buittoni, Gian Luigi, 178
Busso, Giuseppe, 23, 27
Byrne, Rory, 227

Caniato, Alfredo, 17
CAR, 87, 96, 105, 109, 110, 119, 135, 163, 171,
    193, 196, 204, 217, 223, 235
Car & Driver, 87, 177, 212, 225
Carli, Renzo, 42, **43**, **73**, 69, 103
Carmassi, Stefano, 203
Carniglia, Tino, 162
Carraroli, Guglielmo, 40
Carrera Panamericana, **36**, 37, 39
Carrozzeria Fantuzzi, 61
Carrozzeria Ghia, 35, **36**, **47**, 53, 82, 88
Carrozzeria Pininfarina, 40, **41**, **44**, 49, **64**, **66**, 67,
    82, 84, **103**, 103–104, **158**, 197–198, 212,
    215, 228
Carrozzeria Scaglietti, **131**, **132–133**
Carrozzeria Touring, 21, 27, **28**, 68
Carrozzeria Vignale, **30**, 37
Carrozzeria Zagato, 14, 215
Casarini, Giordano, 14
Castellotti, Eugenio, 13
Castriota, Jason, 82, 212–213
Chinetti, Luigi, 31, **35**, 55, 65, **66**, 73, **73**, 82
Chiti, Carlo, **62**, 62, **63**, 65
Cisitalia 202 berlinetta, 40
Citroen, 88
Citroen DS, 199
Cizeta V16T, 154
CMN, 17
Coco, Donato, 228
Cognolato, Dino, **24**, 126, **132–133**, 140
Cognolato, Gino, 118
Coletta, Antonello, 215
Collins, Peter, 13
Colombo, Gioachino, 22–23, 28, 31, 62
Cooper Climax, **62**, 62
Corse Clienti, 217
Cortese, Franco, **23**, 24
Cresto, Sergio, 120
Cropley, Steve, 206
Cunningham, Briggs, 31

Dallara, 178, 182–183
Dallara, Gianpaolo, 71, 74, 79
Dal Monte, Luca, 21
Davenport, John, 119–120
Day, Willametta Keck, 46
de Angelis, Elio, 98
de Angelis, Giuliano, 81, 87
Della Casa, Carlo, 161
De Tomaso, 88
de Vroom, Jan, 55

Dindo, Luigi, 140
Dreyfus, Rene, 17
Ducato, Alfred, 46

Earle, Steve, **64**
Eccelstone, Bernie, 142, 156, 182
Ecurie Francorchamps, 65
Egan, Peter, 177
Evans, Andy, 177

Fangio, Juan Manuel, 13
Farina, Battista "Pinin." *See* Pininfarina, Battista
Federation Internationale du Sport Automobile
    (FISA), 102, 120, 122
Fedeli, Roberto, 225, 229
Felisa, Amedeo, **189**, 189–190, 192–193, 197,
    202–203, 215, 217, 223, 225
Ferrari, Enzo,
    288 GTO and, 100–101
    Anderloni and, 28
    F40 and, 128, 130, 140
    Fiat and, 87–88
    firing of Gardini and, 65
    gap between F1 and road cars and, 100–102,
        101, **102**, 135
    marketing and, 51
    mid-engine configuration and, 62, 65, 71, 81
    Montezemolo and, 190, 192
    personality of, 13–14, 17, 68
    pictured, **14–15**, **22**, **36**, **39**, **43**, **63**, **94**, **128**, **190**
    racing career, 17
    Sergio Pininfarina and, **39**, 40, 42–43, 45, 69,
        71, 74
    World War II and, 21–22
Ferrari, Piero, 35, 37, **155**, 155–158, **156–157**,
    162, 162, **163**, 178, 182, 222–223, **226**
Ferrari Gestione Sportiva, 98, 227
Ferrari-Maserati Design division, **226**, 227–229
Ferrari models,
    33 Prototipo Speciale, 82
    125, **23**, 23–24, 27
    125 C, **29**
    125 replica, **24**
    125 S, **25**
    126 C, 96
    126 C2, 104
    156 F1, **63**
    159, 27
    166, 27
    166 Formula 2, 27
    166 Inter, 27
    166 MM, **27**, 27–29
    166 Sport, **27**, 27
    166 Spyder Corsa, 27
    206/246 Dino, **69**, 81, 103
    208 Turbo, 98
    246 SP, **63**

250 GTO, *60*, 61–62, 154
250 LM, *64*, 65, *66*, 67, *67*, 69, 71
250 LM Speciale, 82, 103
250 P, *64*, 65
250 SWB, 235
275 GTB, *74*
275 GTB/4, 74, *75*, 235
275 P, 73
275 P2, 73
275 S, 31
288 GTO, 98, 100–101, 104–105, *106–109*, 109–110, *112–113*
308 GT/4, 96
308 GTB/GTS, 96, 103
328 GTB/GTS, 98
330 GTC/GTS, 79
330 GTO, 61
330 LMB, *60*, 61–62
330 P/P2, 73
333 SP, 157, *178*, 178, 181
340/375 MM, 37, *38*
340 America, *30*, 31, *32*, *33*, 35, *36*, 37
340 America Vignale Spyder, *35*
340 MM, 37
348, *188*, 192
355, *189*
355F1, *193*, 193
365 GT4 2+2, 103
365 GT4/Berlinetta Boxer, 77, 81–82, 84, *85*, *87*, 87, 103
365 GTB/4 (Daytona Coupe), *76*, 76, *78*, 79–80, 103
365 GTS/4 (Daytona Spyder), 76, *80*, 80
365 P, 71–74, *72*, *73*
375 F1, *29*, *36*, 37
375 MM, 37, 38, *41*, *44*, 45–46, *45–48*, 50
375 Plus, 39, *41*
410 SA, 51–53, *52*, *53*, *55*, 55–56
410 SA Superfast 4.9, *54*, 55
410 SA Superfast I, 53, *53*, 55
410 Series III, *54*, *55*, 56
410 Sport, 49, *51*
410 Super Gilda, *53*, 53, 55
412, 98
412 Cabriolet, *122*, 122, *123*
456 GT, *195*
500 Superfast, 73
512 BB, *88*, *89*
512 BBi, *95*
512 S, 178
512 S show car, *83*
550 Maranello, *196*
599 GTB, 225
612 Scaglietti, *122*
Auto Avio Costruzioni 815, *20*, *21*, 21–22
Birdcage 75 (Maserati), 82, 212–213

Dino Berlinetta GT, 82
Dino Competizione, 82
Dino Speciale, *69*, 71, 82
Enzo Ferrari, 196, 197–199, *199–202*, 202–203, *204*, 205–206, *206–211*, 222
F40, 98, 123, *124*, 125–128, *127–129*, 130–131, *134*, 135, *135–139*, 145–146
F40 GT, 140, 144, 146–147
F40 GTE, 140, 144, 146–147, 148, *148–149*
F40 LM, 140, *141*, 142, *143*
F50, 152, 157, 159–163, *162*, *164–171*, *172–177*
F50 GT/GT1, 157, *180*, 180–183, *182*, *183*
F150, *226*, 227, 229, 231
F512M, *196*
FX, *160*, 162, *186–187*, *193*
FXX/FXX Evoluzione, 215, *216*, 217, *218–219*
GTO Evoluzione, 98, *114–115*, *116*, 116–120, *117*, 122
LaFerrari, *228*, 231, 233, 235
MC12 CORSA (Maserati), 206, 210–211
MC12/MC12 GT1 (Maserati), 206, *210*, 210–211, 213
Millechili static model, 223
Modulo, 82, *83*
Mondial 8, *97*
Mythos, *158*, *159*, 160
P4/5, 213–215
P5, 82, *85*, 103
P6, *82*, 82, 84, 103
Rossa barchetta, 82
Superfast II, 56, *57*, 82
Testarossa, 98, *99*, 110, *150–151*, 154
Ferte, Michel, 142, 145
Fiat, *75*, 88, 94, 98
Fiat Abarth, 98
Fiat Group, 127, 130, 135, 183, 192
Fiat Research Center, 119
Fioravanti, Leonardo, *66*, *67*, 67, 79, *80*, *103*, 103–104, *128*, 159, 194
Fisichella, Giancarlo, 235
Ford, Henry, *70*, 71
Ford GT40, *70*
Ford Motor Company, *70*, 71, 73, 88
Forghieri, Mauro, *63*, 65, 71, 73, 96, *99*, 156
Formula 1 series, 31
FORZA, 189, 225
Fraser, Ian, 96
Frere, Paul, 79, 105, 130–131
Frey, Donald, 71

Gardini, Gerolamo, 40, 49, 65
Garella, Paolo, 212
Gauld, Graham, 61
Gavina, Francesco, 37
Ghidella, Vittorio, 98, 127, 130
Giacomelli, Bruno, 98

Giberti, Frederico, 156
Glickenhaus, Jim, 213
Gobbato, Ugo, 21
Gonzales, Jose Froilan, *36*, 37
Gozzi, Franco, *128*
Green, Gavin, 135, 193
Gross, Ken, 128, 130

Harrah, Bill, 56
Haywood, Hurley, 142
Held, Jean-Francis, 88, 235
Hercules Corporation, 104
Hill, Phil, *63*, 109–110
Hoffman, Paul, 88
Hogan, John, 192
Honda Acura NSX, 225
Hong, Patrick, 206
Horrell, Paul, 171, 193

Iacocca, Lee, 71
IMSA, 142, *143*, 145, 157, 178
J. S. Inskip Rolls Royce, 31
Iso, 74, 88, 154
   Grifo, 74, *80*, 154
   Grifo 7 Liter, 80
Isuzu 4200R, 154

Jaguar,
   C-X 75, 225
   XJ220, 154, 163
   XK120, 35

Kimberly, Jim, 39
Klein, Reinhard, 119–120

Laffite, Jacques, 142
Lamborghini, 88
   Countach, 80, *84*, 110, *152*, 154
   Miura, 71, 74, 79–80
   P400, *69*, *70*, 71
Lamm, John, 128, *128*, 129
Lampredi, Aurelio, 31, *36*, 37
Lancia,
   Montecarlo Turbo, 125
   Stratos, 96, *99*, 116
   Stratos Silhouette, 96
Lancia Aprilia Berlinetta Aerodinamica, 40
Lancia Corse, 98
Larini, Nicola, 182
Lauda, Niki, *162*, 163, 192
Lawrence, Brandon, 109, 154
Lee, Tommy, 31
Le Mans, 29, 39, 65, 72, 140, 142, 148, 157
Leopold, King of Belgium, *48*, 50
Lini, Franco, 40

Macchiavelli, Lotario Rangoni, 21

MacKenzie, Angus, 190

Manfredini, Maurizio, 126

Mansel, Nigel, 158

Manton, David, 21–22

Manzoni, Flavio, *226*, 228–229

Maranello facility, 22, 24, 65, *94*, *191*

Marcus, Frank, 211

Marin, Daniel, 140, 142

Marshall, George, 29

Martinengo, Franco, *44*, 46, 55

Marzotto, Giannino, 31

Maserati, 24, 29, 31, 88
  5000 GT, 56
  A6, 40
  Birdcage 75, 82, 212–213
  Ghibli, 74, *75*, 80
  MC12 CORSA, 206, 210–211
  MC12/MC12 GT1, 206, *210*, 210–211, 213

Maserati, Alfieri, 17

Massimino, Alberto, 21

Massini, Marcel, 37

Materazzi, Nicola, 96, 98–99, *99*, 100–102, 105, *114–115*, 116–119, 122, 125, *128*, 130, 135, 140, 154

McLaren, 104, 147, 227
  F1, 73, 104, 154, *155*, 163, 205–206, 222
  GTR, 147
  MP4/1, 104
  MP4-12C, 227
  P1, 227, 233

Meiners, Franco, 144, 145–146, 180, 182–183

Mercedes 300 SL Gullwing, 53

Michelotti, Giovanni, 37

Michelotto, Giuliano, 118, 140, 145, 146, 178

Migliore, Greg, 233

Mille Miglia, *16*, 21, 29, 37

Modena, 21, 28

Modena Aeroautodromo, *63*

MOMO, 157, 178

Moncalesi, Maurizio, *203*, 211, 217

Monteverdi,
  400 SS, 80
  Hai, 80, *81*

Montezemolo, Luca Cordero di, 156, 157, *163*, 178, 189–190, *190*, 192–193, 197–199, *198*, *226*, *228*, 231

Moretti, Gianpiero, 157, 178, *179*

Moro, Aldo, 88

Mosely, Max, *223*, 223

Moss, Stirling, *36*, *62*, 62

The Motor, 46

Motor, 79

Motorsport, 109

Motor Trend, 190, 193, 211

Murray, Gordon, 147

Musso, Luigi, 13

Nardi, Enrico, 21, 22

NASCAR, 157

Neilson, John, 147

Nichols, Mel, 87

Nicodemi, Adriano, 145

Okuyama, Ken, 199

Olofsson, Anders, 147

OSCA, 29

Osella, 96

Packard, 23

Parkes, Mike, 14, 61, *73*

Paul-Cavallier, Michel, 46, 49

Pegaso, 53

Peroni, Sergio, 145

Petrotta, Giuseppe, 197, 202, 217

Piccinini, Marco, 156

Pininfarina, Andrea, 212, 228

Pininfarina, Battista, 29, *39*, 40, *44*, 49, *74*

Pininfarina, Sergio, 13, *39*, 40, 42–43, *43*, *44*, 45, 49, 51, 68, 69, 71, *73*, 74, *77*, 85, 88–89, 94, 103, *159*, *162*, *163*, 197, *198*, 199, 228

Pinto, Raffaele "Lele," 119

Ponzoni, Gaetano, 28, 68

Porsche,
  911, 120
  918 Spyder, *224*, 233
  930, 116
  959, 119, 122, 130–131, 135
  962, 131
  Gruppe B, *121*, 122

Postlethwaite, Harvey, 104

Pourret, Jess, 61

Pozzi, Charles, 140

Prost, Alain, 158

Purdy, Ken, 13

Quattroruote, 67

Ramaciotti, Lorenzo, 103, 125–126, 159–161, *159*, 189, 192, 194, *195*, 198–199, 203, 235

Razelli, Giovanni, 127, *128*, 140, 154

Renault, 96

Renault 18 Turbo, 102

Ricart, Wifredo, 21

Rivolta, Piero, 51, 74, 88, 154

Road & Track, 35, 55, 56, 71, 82, 84, 105, 109, 110, 128–130, 177, 205, 217

Rocchi, Franco, 96

Rossellini, Roberto, 46, 49

Rudd, Bill, 56

Sage, Jean, 142

Salvarani, Walter, 96

Santos, Joaquim, 120

Sapino, Filippo, 82

Savonuzzi, Giovanni, 53, 55

Scaglietti, Sergio, 14, *15*, 42, *45*, *46*, 46, 49, 61, 68, *75*, *80*, 87, *122*, 122, *123*, *190*

Scannavini, Michele, 163

Schell, Lucy, 31

Schumacher, Michael, 182, 198, 203

Scuderia Ferrari, *15*, *16*, 17, 21

Scuderia Serenissima, 13

Segre, Luigi, 35

Sguazzini, Giovanni, 118

Shelby, Carroll, 13

Sheriff, Antony, 227

Silenzi, Claudio, 225

Silverstone Grand Prix, 37

Simanaitis, Dennis, 217

Societa Auto Avio Costruzioni, 21

Sports Car Club of America (SCCA), 29

Sports Car Club of California, 29

Sports Car Illustrated, 51, 55–56

Stanguellini, 29

Stephenson, Frank, 211, 227–228

Stuck, Hans, 142

Studebaker Land Cruiser, 35

Subaru Jiotto Caspita, 154

Swaters, Jacques, 65

Tavoni, Romolo, 65

Tjaarda, Tom, 88, 94

Toivonen, Henri, 120

Toni, Marco, 122

Trintignant, Maurice, *62*, 62

Turin, 88

Varzi, Achilli, *17*

Vernor, Brenda, 14, 17

Vignale, Alfredo, 37

Villoresi, Gigi, *32*, 37

Volpi, Giovanni, 13, 40

Von Trips, Wolfgang, *63*

Wallace, Bob, *84*

Walton, Mark, 205

Watson, John, 104

Wax, Enrico, 46

Whitmarsh, Martin, 190

Wilke, Robert C., *47*, 49

World Championship Rally (WRC), 102

Zafiropolo, Art, 183